*Advances in Polymer Blends
and Alloys Technology—Volume 2*

T0174822

Advances in Polymer Blends and Alloys Technology

VOLUME 2

Edited by Melvyn A. Kohudic and Kier Finlayson

CRC Press
Taylor & Francis Group
Boca Raton London New York

CRC Press is an imprint of the
Taylor & Francis Group, an **informa** business

CRC Press
Taylor & Francis Group
6000 Broken Sound Parkway NW, Suite 300
Boca Raton, FL 33487-2742

© 1989 by Taylor & Francis Group, LLC
CRC Press is an imprint of Taylor & Francis Group, an Informa business

First issued in paperback 2019

No claim to original U.S. Government works

ISBN 13: 978-0-367-45096-0 (pbk)
ISBN 13: 978-0-87762-670-1 (hbk)

**Visit the Taylor & Francis Web site at
http://www.taylorandfrancis.com**

**and the CRC Press Web site at
http://www.crcpress.com**

TABLE OF CONTENTS

1

Mechanical Behaviour of Blends of Polystyrene with Styrene-Isoprene-Styrene (SIS) and Ethylene-Propylene Rubber (EPR)

NEFAA MEKHILEF,* BUNYAMIN ELBIRLI* and TOUFIK BAOUZ*

INTRODUCTION

IT IS A well-known fact that the incorporation of a rubbery phase as a modifier can improve the impact strength of glassy polymers, notably that of polystyrene. However, the impact strength obtained depends on both the amount of rubber incorporated and the method of forming the blend. The amount of rubber incorporated should be such that a high impact strength is obtained without adversely affecting other mechanical properties such as tensile, flexural strength, etc., of the glassy polymer. The method of forming the blend is also important, since it will affect the final blend morphology and, therefore, the size of the rubbery domains as well as the mechanical nature of the interphase [1,2]. There are various methods of forming a matrix, such as: melt blending, graft copolymerization, and interpenetrating networks (IPN). The rubbery domains serve as stress concentrators that can induce crazing which contributes considerably to the dissipation of impact and strain energy.

The role of the rubbery domains is not only to initiate crazes, but they should also share the load with the matrix after crazing has occurred. Another requirement for the elastomeric phase is that its T_g must be much lower than the test temperature [2].

The objectives of the present study are:

(a) To modify polystyrene with a linear thermoplastic elastomer of styrene-

* Department of Plastics & Rubber Engineering, Algerian Petroleum Institute (I.A.P.), 35000 Boumerdes, Algeria.

1

isoprene-styrene triblock copolymer (SIS) up to 20% by weight by melt blending

(b) To modify polystyrene with ethylene-propylene rubber (EPR) up to 8.8% by weight by melt blending, and to compare the resulting morphologies and mechanical properties with those available for a chemically prepared blend such as high impact polystyrene (HIPS). Although blends of SBS (styrene-butadiene-styrene) or SB (styrene-butadiene) with PS have been studied previously [3,4,5], there is not much information in the literature concerning PS/SIS blends. SIS, mechanically and chemically, is very different compared to its SBS counterpart. Compared to Kraton 1101 (SBS), Kraton 1107 (SIS) has a lower weight percentage of PS and therefore exhibits lower specific gravity and tensile strength but higher elongation at break [4]. Similar to SBS, SIS is a two-phase system with PS blocks having a molecular weight of approximately 7000 and forming spherical domains of 400 Å.

Figure 1(a) shows the simplified schematic representation of the linear SIS block copolymer. In PS/SIS blends, SIS forms the minor phase, retaining its two-phase morphology, and the homopolymer PS forms the continuous third phase.

Figure 1(b) schematically illustrates the three-phase morphology and the polyisoprene/PS interphase where the situation is similar to an emulsion with PS blocks segregating along the phase boundaries. PS/EPR blends, however, are two-phase systems with no apparent interaction at the interphase [Figure 1(c)]. In this work, the following mechanical properties were investigated for the PS/SIS and PS/EPR blends:

- notched and reversed notched Izod impact strength
- tensile strength at yield and break
- elongation at break

Additionally, blend morphology was examined by scanning electron microscopy.

EXPERIMENTAL

Specimen Preparation

Because standard thermoplastic processing techniques (extrusion, injection moulding) are practical and economical, they were used to prepare the blends. The following materials were used to perform this study: glassy thermoplastic, a general purpose PS (Dow Chemical) for the PS/SIS blend and a general purpose PS (BASF) for the PS/EPR blend; thermoplastic elastomer, styrene-isoprene-styrene (SIS) triblock copolymer (Shell Kraton 1107) PI/PS, 68/14; ethylene-propylene rubber (EPR), Exxon Chemicals with no diene content; linear copolymer, Vistalon 917; and high impact polystyrene, Lacqrene (HIPS) ATOCHEM.

Blends containing 2, 5, 10, 20, and 30% by weight of SIS in PS and 2, 5, and 8.8% by weight of EPR in PS were prepared in a ¾ " laboratory extruder operating at 200°C and 80 rpm. Blends obtained as rods in the first case were granulated and passed through the extruder a second time.

Tensile and impact test specimens of the two blends and HIPS were prepared

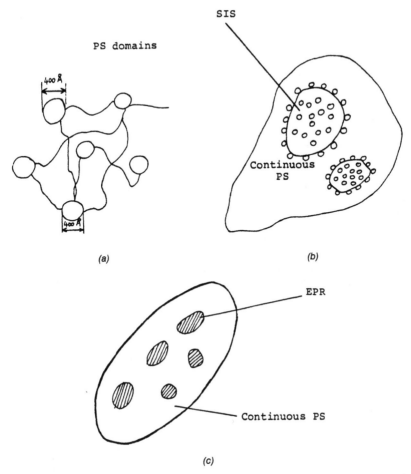

FIGURE 1. (a) Schematic representation of a linear SIS block copolymer, (b) three-phase morphology of PS/SIS blends, and (c) two-phase morphology of PS/EPR blends.

using an injection moulding machine operating under the following conditions: injection temperature, 200°C; injection pressure, approximately 1200 kg/cm²; injection time, 1 sec; injection rate, 26 cm³/sec; and cooling time, 20 sec.

Mechanical Tests

The Izod impact strength of notched PS/SIS and PS/EPR blends and HIPS was determined according to ASTM D256, and the results were reported in both kJ/m² of broken surface and the more familiar lb$_f$-ft/in of notch. A reversed notch Izod impact test was also carried out to provide a measure of the unnotched

impact strength. The tensile tests were performed on the Instron machine according to ASTM D638, and the specimens were deformed at two different crosshead
speeds: 20 and 200 mm/min for PS/SIS specimens and 20 mm/min for PS/EPR
specimens. The stresses at yield (σy) and at break (σb) were determined and
expressed in MPa and in psi, and the elongation at break (ϵb) in %. Ten specimens of each blend were evaluated for each test and their mean values are given
as a function of the blend composition expressed as % SIS, % EPR (modifier),
and also as % diene rubber-content for the PS/SIS blend.

Electron Microscopy

Specimens broken at room temperature were examined using a Bausch and
Lomb scanning electron microscope. This was done in order to examine the morphology of the broken surfaces of the blends at relatively high magnifications
(above × 24,000).

RESULTS AND DISCUSSION

Izod impact strength of notched specimens is shown in Figure 2, where a
gradual increase in the impact strength is observed as the SIS content increases.

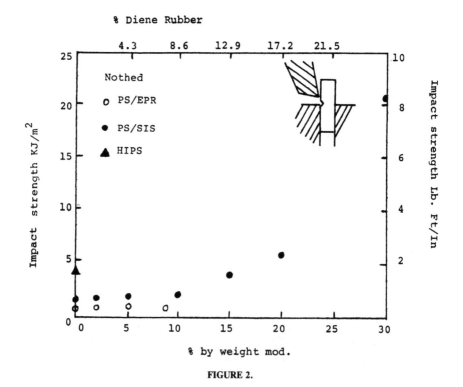

FIGURE 2.

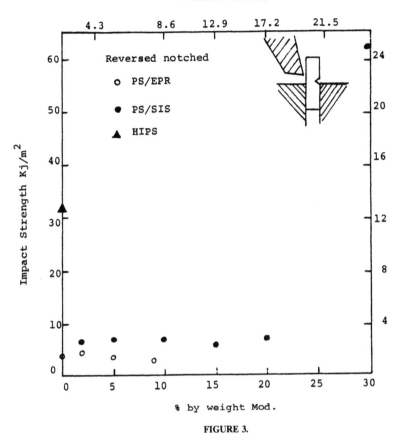

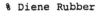

FIGURE 3.

The impact values for PS/EPR blends are comparatively lower than PS/SIS blends which is attributed to the poorer interaction between the modifier and matrix. The impact strength of the reversed notched specimens shown in Figure 3 remains fairly constant up to 20% of SIS. Similar to Figure 2, PS/EPR blends exhibit lower values. The inclusion of SIS above 20% by weight resulted in considerable softening and a drastic increase in the impact strength. The notched samples exhibited lower impact strength compared to unnotched samples, and this indicates that the blends as well as HIPS are highly notch sensitive. The notch offers the opportunity to initiate crazes faster and develop them into a catastrophic crack, and therefore, it renders the blends less sensitive to the state of dispersion of the rubbery phase. Except for the notched specimens containing 15% SIS and the unnotched specimens containing 25% SIS, the results of the Izod impact test for both the notched and unnotched samples of blends did not

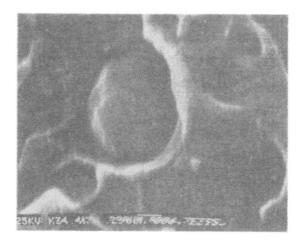

FIGURE 4a. PS/EPR 8.8% (EPR Vistalon 719, no diene content, Exxon Chemicals; PS general purpose, BASF).

quite reach those of HIPS. At large magnifications, the SEM picture Figure 4a revealed the formation of voids with sharp boundaries around EPR particles. Upon cooling, a very weak interphase results because of the differential shrinkage between the EPR particles and the PS matrix. Figure 4b shows the SEM picture of the fracture surface of a blend containing 10% SIS in PS. The surface morphology reveals numerous voids that were presumed to have previously been

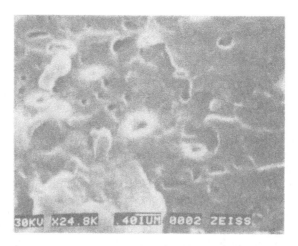

FIGURE 4b. PS/SIS 10% (SIS Kraton 1107, Shell; PS general purpose, Dow).

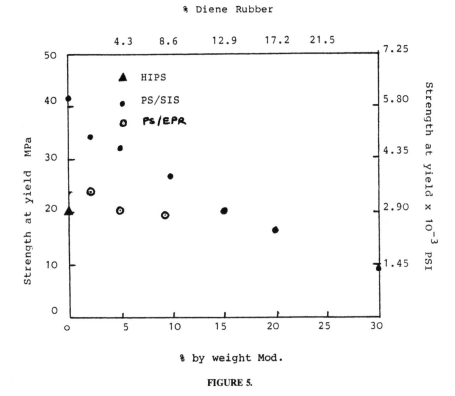

FIGURE 5.

occupied by the SIS phase. At 24,800 × magnification it becomes clear that these voids cannot be attributed to the volatiles, since voids formed by volatiles are visible even to the naked eye. Figure 4b shows that the boundaries of the voids are regular, thus indicating a weak interface between SIS and PS. This could be attributed to:

(a) The incompatibility of low molecular weight PS blocks in SIS with the commercial PS of much higher molecular weight
(b) The weak interface between PS domains and PI phase in SIS, as reported in the literature [4]

Thus, the morphology of the blends seems to confirm the Izod impact strength results for the PS/SIS and the PS/EPR blends.

Tensile Tests

The results of tensile strength at yield and break for HIPS, PS, PS/SIS, and PS/EPR blends are shown in Figures 5 and 6. It is observed that the strength of the blends decreases as the content of rubber increases. In the range of 2–15% by

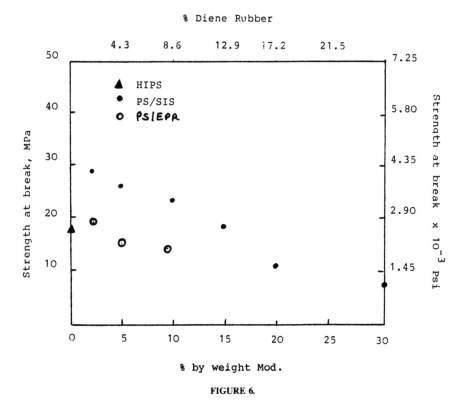

FIGURE 6.

weight of SIS, the blends exhibit higher strength compared to HIPS alone. At low SIS content, however, the tensile strength of PS is retained. When the SIS content is greater than 15%, the tensile strength of the blends decreases significantly below the strength observed for HIPS alone. When the specimens were deformed at relatively high speed, no significant change in the strength of the blends and HIPS was observed within the normal scatter. The tensile strength for PS/EPR blends decreases as the EPR content increases. The strength values of these blends are lower compared to those of PS/SIS and are attributed to:

(a) Slightly inferior mechanical properties of the PS phase
(b) The presence of voids between PS and EPR particles

Table 1 summarizes the results of tensile strength for the PS/SIS blends and HIPS when deformed at 20 and 200 mm/min. Table 2 shows the results of tensile strength for the PS/EPR blends when deformed at 20 mm/min, as previously reported. The results indicate that, within the range of strain rates employed for the PS/SIS blends, the deformation mechanism (initiation of crazes and local yielding of block copolymer particles [3]) remains unchanged.

Table 1.

SIS %	σy (MPa)[a]	σy (MPa)[b]	σb (MPa)[a]	σb (MPa)[b]	ϵb %[a]	ϵb %[b]
2	34.2	34.6	29.1	32.0	8.4	8.6
5	32.2	34.5	26.1	35.5	9.0	10.8
10	26.5	23.3	23.8	19.0	3.8	3.8
15	19.5	15.0	18.8	11.7	3.1	3.4
20	16.5	14.5	11.5	11.9	4.3	3.4
30	9.5	11.0	8.2	10.2	16.2	25.5
HIPS	20.0	19.9	18.6	19.8	45.6	41.6

[a]Deformation rate: 20 mm/min.
[b]Deformation rate: 200 mm/min.

Elongation at Break

The elongation at break at the 20 mm/min strain rate is given in Figure 7, and Table 1 shows that the results for the PS/SIS blends are practically independent of the strain rate employed. Compared to PS alone, at relatively low SIS contents, the blends exhibit somewhat higher elongation. The elongation at break is seen to be quite low for blends containing 10–20% SIS; at high SIS content, however, the elongation increases dramatically. Similarly, the elongation at break of PS/EPR blends is much lower than that of HIPS and greater than that of PS alone. However, it shows a uniform decrease as the EPR content increases. HIPS shows the highest elongation at break. These results seem to reinforce the observation of the weak interphase between SIS and PS and EPR and PS.

CONCLUSIONS

- The relatively poor impact strength and elongation at break at low SIS or EPR contents can be attributed to the weak interphase between the SIS or EPR and the continuous polystyrene phase.
- Strength at yield and break for the blends decreases as the modifier content increases.

Table 2.

EPR %	σy (MPa)	σb (MPa)	ϵb (%)
2	24.0	18.9	10.8
5	20.2	15.5	9.1
8.8	19.5	14.5	14.9
PS	—	27.2	2.4

Deformation at break: 20 mm/min.

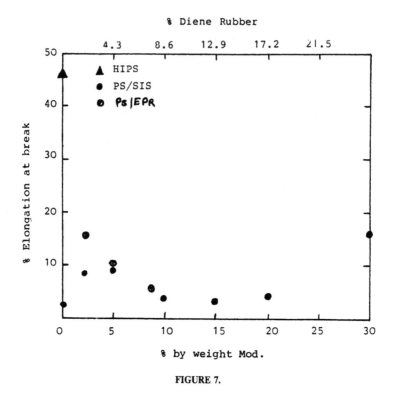

FIGURE 7.

- Voids exist between PS and EPR particles.
- For PS/SIS blends, strength at yield, strength at break, and elongation at break do not depend on the rate of deformation.
- At high SIS content (20%), where the amount of rubber in blends approaches the level of rubber in commercial HIPS products, the notched Izod impact strength of the blends compares well with that of HIPS. However, SIS content above 10% by weight in PS is not desirable in terms of tensile strength.

ACKNOWLEDGEMENT

We appreciate Ms. Habri of Industrial Chemistry at IAP for her help in preparing the scanning electron micrographs.

REFERENCES

1. Kresge, E. N. "Elastomeric Blends," *J. of Appl. Polym. Sci.*, Applied Polymer Symposium 39, 37 (1984).

2. Manson, J. A. and L. H. Sperling. *Polymer Blends and Composites.* New York:Plenum Press (1976).

3. Folkes, M. J. *Processing, Structure and Properties of Block Copolymers.* London:Elsevier Applied Science Publishers, Chapter 1 (1985).

4. Thorn, A. D. *Thermoplastic Elastomers.* Rubber and Plastics Research Association of Great Britain (1980).

5. Swisher, G. M. and R. D. Mathis. R.D. 186 ANTEC (1984).

2

Effect of the Molecular Structure of *in situ*-Generated Ethylene-Propylene Copolymers on the Impact Properties of Polypropylene

WILLIAM D. VERNON*

INTRODUCTION

ISOTACTIC POLYPROPYLENE IS not intrinsically tough at lower temperatures, yet its other properties make it attractive for applications where toughness is desirable. Polypropylene is rendered impact resistant when ethylene-propylene copolymeric elastomers are incorporated, commonly at the 5–15% level. This can be done either by blending or by synthesizing ethylene-propylene elastomers *in situ* in polypropylene homopolymers.

Production of *in situ* blends can be accomplished by several methods. A simple example is the production of isotactic polypropylene in the first stage of the reactor which is then conveyed, along with the still-active Ziegler-Natta catalyst, to a second stage into which ethylene and propylene monomers are introduced. There a noncrystalline, elastomeric copolymer is made which is blended intimately with the polypropylene powder. A small amount of hydrogen, present as a chain-transfer agent in the first stage, is swept into the second stage along with the polypropylene powder. The mechanical transfer of powder requires several seconds—many times longer than the chain propagation lifetime. Thus, the final product is an intimate mixture of polypropylene homopolymer and ethylene-propylene elastomer. Active polymerization sites make their way into the second reactor, but any "living chains" are terminated by hydrogen before the homopolymer enters the second stage.

Through the years of manufacturing impact-resistant polypropylenes we have

---

* El Paso Products Company, Odessa, Texas.

been puzzled by the frequent observation that different lots of resin containing about the same levels of ethylene, as measured by infrared spectroscopy, exhibit widely disparate impact properties. Plant wisdom dictated that to increase impact strengths, it was necessary to incorporate more ethylene in the product, though we often observed that increased ethylene levels did not necessarily result in increased impact strengths. To explain these observations, we reasoned that perhaps the manner in which ethylene is incorporated in a resin, not simply the gross quantity of ethylene, determines the resin's impact strength. Therefore, we undertook a study to determine how ethylene is incorporated in impact-resistant polypropylene resins and how its incorporation correlates with impact properties.

EXPERIMENTAL

Impact-resistant resins were fractionated in xylene by a method described by Simonazzi [1]. Five grams of polymer were dissolved in 1000 ml of boiling *o*-xylene. The solution was allowed to stand at ambient temperature 24 hr, undisturbed, during which time a flocculent precipitate formed. The precipitate was filtered, washed with acetone, and dried *in vacuo* at 60°C. The end result was a white, crystalline powder. The solvent was removed from the filtrate on a rotary evaporator, and the resulting rubbery residue was scraped from the walls of the flask, washed with acetone, and similarly dried *in vacuo*. The precipitate typically comprised about 85% of the original sample, the elastomer the remaining 15%. These will be referred to as the insoluble and soluble fractions, respectively.

High-resolution ^{13}C NMR spectra were measured at 67.80 MHz on a JEOL FX270 spectrometer with complete proton decoupling at 269.65 MHz. Five thousand transients were collected; a pulse interval of 15 sec was employed to ensure complete relaxation of the nuclear spins. Pulse widths of 9 μsec gave a tilt angle of ca 70°. The probe temperature was maintained at 125°C to ensure complete solubility of the sample. Solutions were prepared by placing ca 250 mg of the sample in a 10 mm NMR tube then adding ca 3 ml of 90%/10% v/v 1,2,4-trichlorobenzene/perdeuterobenzene, the latter added to serve as an internal deuterium lock. A few mg of *N*-phenyl-1-naphthylamine were added to protect the solution from oxidation. The NMR tube was then placed in an oil bath at 130°C, whereupon the polymer melted and dissolved and the solution was degassed. The solution was then transferred to the spectrometer. Comonomer contents of both fractions and multad frequencies of the soluble fractions were calculated from the integrated intensities of the resonances following the methods outlined by Ray et al. [2].

DSC scans were measured on a Perkin-Elmer DSC-2C calorimeter equipped with a 3600 data station. Prior to measuring the scans, we annealed the samples by heating them to 225°C at 80°C/min, holding them at that temperature for 5 min, cooling them to 150°C at 320°C/min, and finally cooling them to 50°C at a rate of 10°C/min. Melting points were then measured using a scan rate of 10°/min.

Notched Izod impact strengths were measured at 23°C according to ASTM

Test D256, Method A. Gardner impact strengths were measured at −40°C according to ASTM Test D3029, Method G.

RESULTS

Initially 15 impact-resistant polypropylene samples from a pilot plant were analyzed for this study. Pilot-plant-generated samples were used because:

1. They had all been made with the same catalyst system under the same conditions.
2. The ethylene levels spanned a broad range.
3. The melt-flow rates were very close.
4. The impact strengths varied considerably.

Figures 1 and 2 show Gardner and Izod impact strengths plotted against total ethylene content. The points are scattered randomly; linear regression yielded a correlation coefficient of −0.10 for the Izod values and 0.03 for the Gardner values. Counter to conventional wisdom the impact resistance of these resins clearly does not correlate with gross ethylene content. We reasoned that the manner in which ethylene is incorporated in the resin might then determine the impact properties.

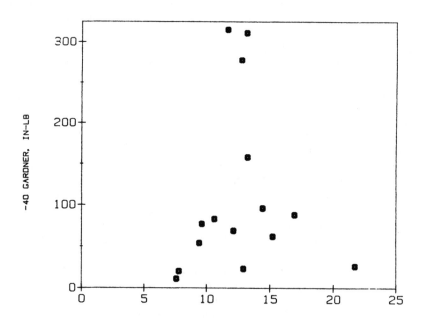

TOTAL E, MOLE PER CENT

FIGURE 1.

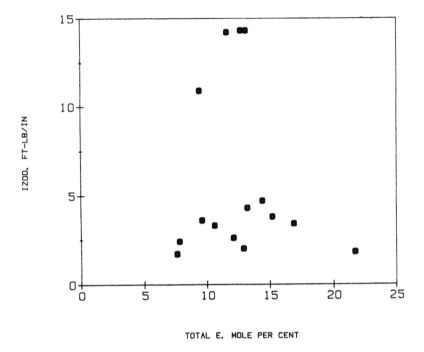

FIGURE 2.

The samples were fractionated in xylene, and the resulting fractions were analyzed by NMR and DSC. The ^{13}C NMR spectrum of a typical xylene-insoluble fraction is shown in Figure 3. It shows the three resonances characteristic of isotactic polypropylene and a resonance due to the presence of a small amount of polyethylene. No ethylene-to-propylene linkages were observed, as evinced by the absence of $S\alpha\delta$ and $S\alpha\gamma$ resonances. Polyethylene levels as determined by NMR usually ranged from 2% to 5% by weight. DSC scans showed endotherms from polypropylene at peak temperatures of 160–165°C and from polyethylene at peak temperatures of 115–120°C. Figure 4 shows the DSC scan of a xylene-insoluble fraction.

The xylene-soluble fractions are rubbery materials which yielded no detectable endotherms when analyzed by DSC, indicating that they are amorphous. Figure 5 shows an NMR spectrum typical of the xylene-soluble fractions. It shows that the xylene-soluble fractions consist primarily of an ethylene-propylene copolymer. The chemical shifts of the $S\alpha\gamma$ and $S\alpha\delta$ resonances, which arise from P-E linkages, indicate that the polypropylene sequences in the copolymer are *meso*; there are no detectable racemic $S\alpha\gamma$ and $S\alpha\delta$ resonances [3]. The absence of an $S\alpha\beta$ resonance indicates there is no head-to-head or tail-to-tail propylene addition in the copolymer.

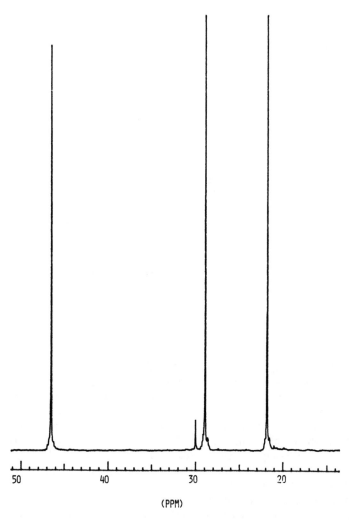

FIGURE 3. ^{13}C NMR spectrum of a xylene-insoluble fraction.

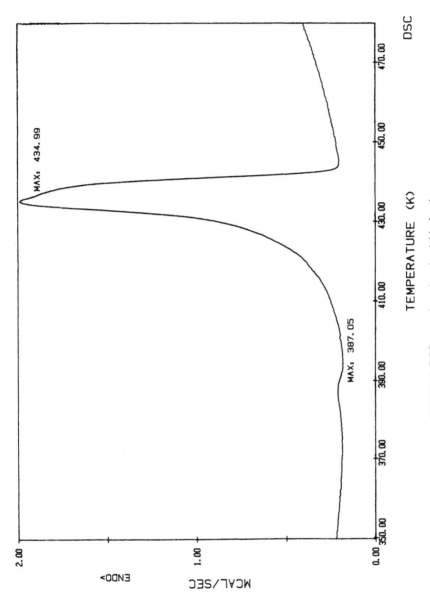

FIGURE 4. DSC scan of a xylene-insoluble fraction.

17

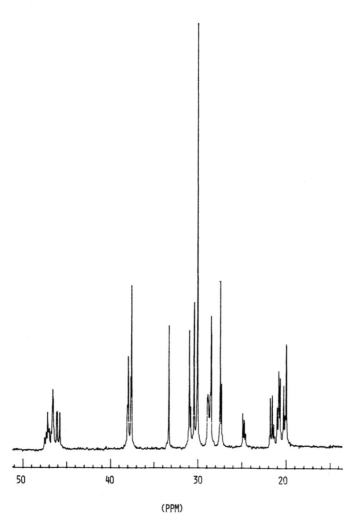

(PPM)

FIGURE 5. ^{13}C NMR spectrum of a xylene-soluble fraction.

18

We learned from previous studies of polypropylene homopolymers that a small amount of lower-molecular-weight homopolymer remains dissolved in xylene at room temperature. Therefore, we surmised that the xylene-soluble fractions of the samples involved in this study contain a small quantity of polypropylene homopolymer as well as the predominant ethylene-propylene copolymer. We made no attempt to remove it, though, since such efforts invariably prove to be fruitless [4]. We realize, too, that the ethylene-propylene copolymer itself comprises a complex mixture of copolymers, but for this study we elected to characterize the soluble fractions by their average structures, as though they were pure components. The ethylene levels of the soluble fractions ranged from 32% to 48% by weight.

From the integrated intensities of the resonances we calculated the distribution of ethylene and propylene groups in the xylene-soluble fractions, in terms of multads, as though the samples were monolithic. The multads of interest were the monads P and E, the dyads PP, PE, and EE, and the triads PPP, PPE, EPE, PEP, PEE, and EEE. We also calculated the quantities E_1, E_2, and E_{3+}, which represent ethylene sequence distributions; that is, they represent the number of moles of ethylene in a given sequence length divided by the total moles of ethylene in the soluble fraction. Thus E_1 is the fraction of ethylene present in the polymer as

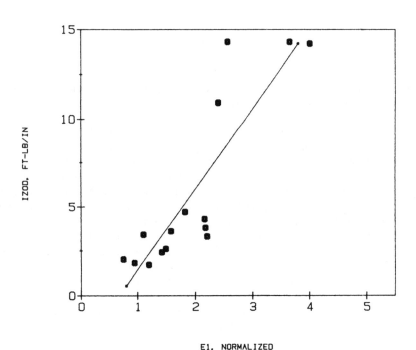

E1, NORMALIZED

FIGURE 6.

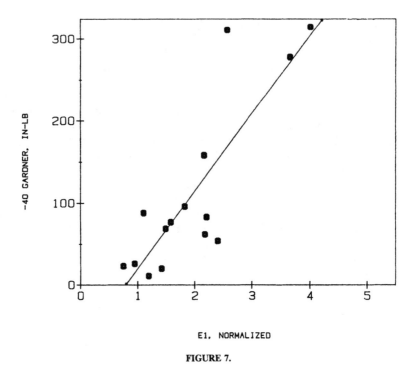

E1, NORMALIZED

FIGURE 7.

isolated, single units, E_2 is the fraction of ethylene present as isolated EE units, and E_{3+} represents the ethylene present in ethylene chains three or more units long [5].

We then plotted Izod and Gardner impact strengths against the frequency of occurrence of each of the multads and against the ethylene sequence distributions, but in no case did we observe significant correlations. However, we obtained significant correlations when we plotted the impact strengths against several of these quantities after the latter had been normalized. Normalization was accomplished by multiplying each multad or sequence-distribution frequency by the percent solubility of the original sample in xylene. For instance, if the PPE triad comprised 15% of the copolymer fraction and the resin was 15% soluble in xylene, the normalized PPE frequency would be 15% × 15%, or 2.25%. Thus, the normalized structural parameters represent the molar concentration of each parameter with respect to the original, unfractionated sample.

Figures 6 and 7 show the plots of Izod and Gardner impact strengths against normalized E_1—the fraction of ethylene present as isolated units in the copolymer fraction. In both cases, linear regression yielded correlation coefficients of approximately 0.85. However, it should also be pointed out that the plots can be interpreted in terms of a step function. A colleague pointed out that in Figure 6, at a normalized E_1 value of about 2.4%, the Izod impact strength suddenly leaps

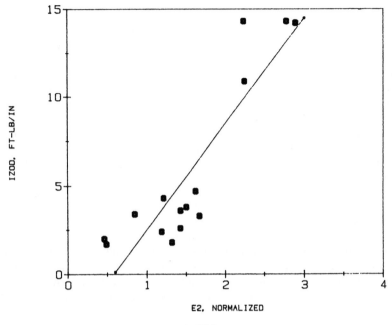

FIGURE 8.

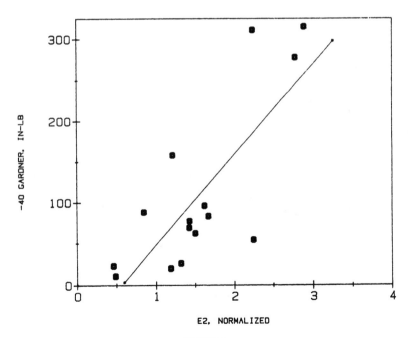

FIGURE 9.

to much higher values and then apparently levels off. Virtually the same trend can be seen in the case of Gardner impact strengths in Figure 7.

Figures 8 and 9 show similar plots of Izod and Gardner impact strengths against normalized E_2. The linear correlation coefficient for the Izod plot is 0.90; it is lower for the Gardner plot, being only 0.78. However, the step function effect can be seen in these plots as well. The leap occurs in the neighborhood of 2.0%.

Plots of Izod and Gardner impact strengths against E_{3+} produced randomly scattered points which gave neither acceptable linear correlation coefficients nor apparent step functions. Thus, we find that both Izod and Gardner impact strengths improve as the amount of ethylene tied up in only the shortest chains is increased. One might predict that ethylene in uninterrupted sequences only three or four ethylene units long might also contribute positively to Izod and Gardner impact strengths, but this cannot be proven by NMR which, unfortunately, is incapable of discriminating between a sequence only three ethylene units long and one much longer.

In order to find other relationships, we plotted Izod and Gardner impact strengths against all the normalized multad frequencies and found acceptable fits when we plotted the multads PEP and PPE. Figures 10 and 11 show the plots involving PEP—the multad that explicitly features an isolated ethylene unit. The linear correlation coefficients in both cases are 0.82. The step effect shows up at

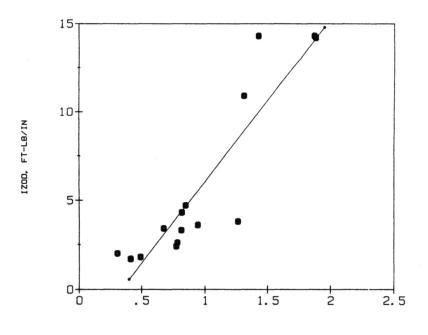

PEP. NORMALIZED

FIGURE 10.

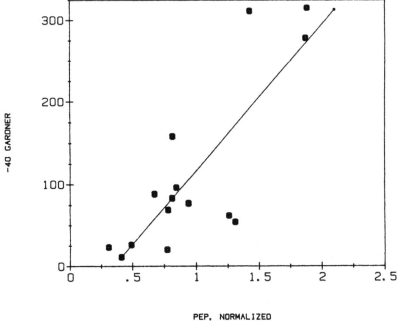

PEP. NORMALIZED

FIGURE 11.

a normalized PEP value of ca 1.3%. The fit with PPE was somewhat unexpected, since, more often than not, the terminal E would be expected to be part of a longer polyethylene sequence. However, the fit is there. The linear correlation coefficient is 0.86 for the Izod impact strengths and 0.84 for the Gardner impact strengths, and in both cases, the step effect begins around 2.3% normalized PPE.

Interestingly, no fit was observed with the normalized monad E, which represents the ethylene content of the soluble fraction. Nor did we observe any kind of fit with the normalized, ethylene-containing multads EE and EEE. All three of these multads are more frequently than not tied up in even longer ethylene chains, so perhaps in retrospect, this lack of correlation should not be surprising.

Earlier it was stated that the xylene-insoluble fractions comprise crystalline polyethylene on the order of 2–5%. We plotted the Izod and Gardner impact strengths of the resins against the quantity of ethylene in the xylene-insoluble fractions and found neither a significant linear correlation nor a step function relationship. We found that the crystalline polyethylene component does not affect the impact properties of the parent resin in any systematic fashion.

CONCLUSIONS

This study has shown that the impact resistance of polypropylene toughened by an *in situ*-generated ethylene–propylene elastomer, as measured by notched Izod impact strength at ambient temperature and by Gardner impact strength at $-40°C$, cannot be related to the gross ethylene content of an intact sample. Instead, the impact properties of such resins depend on the quantity and average structure of the copolymeric fraction. The results indicate that ethylene present in the elastomer in the shortest sequences, one or two ethylene units in length, contributes most highly to impact resistance. Neither ethylene incorporated in longer polyethylene sequences in the elastomer, nor ethylene present as crystalline polyethylene, appreciably contribute to impact resistance. Maximum impact resistance is observed in resins whose copolymer fractions have ethylene concentrated in the shorter polyethylene moieties.

Presently we are attempting to understand these observations. The mechanism by which an elastomer such as the ethylene–propylene copolymer imparts impact resistance to crystalline propylene has been studied extensively. Studies have shown that such mixtures are two-phase systems; consequently there are domains of elastomer scattered randomly throughout the crystalline polypropylene continuous phase [6,7]. Other studies have concluded that resins with smaller, more highly dispersed elastomer domains have superior impact resistance [8]. We hypothesize that a predominance of shorter polyethylene sequences in the elastomer fraction modifies its blending characteristics in a manner favorable to enhanced impact resistance. We are presently attempting to correlate the observations discussed herein with features observed in photomicrographs of the resins to verify this theory.

REFERENCES

1. Simonazzi, T. *Pure & Appl. Chem.*, 56:625 (1984).
2. Ray, G. J., P. E. Johnson and J. R. Knox. *Macromolecules,* 10:773 (1977).
3. Cheng, H. N. *Macromolecules,* 17:1950 (1984).
4. Ogawa, T. and T. Inaba. *J. Polym. Sci., Polym. Phys. Ed.,* 12:785 (1974).
5. Cheng, H. N. *Anal. Chem.,* 54:1828 (1982).
6. Karger-Kocsis, J., L. Kiss and V. N. Kuleznev. *Polym. Commun.,* 25:122 (1984).
7. Kalfoglou, N. K. *Angew. Makromol. Chem.,* 129:103 (1985).
8. Stehling, F. C., T. Huff, C. S. Speed and G. Wissler. *J. Appl. Polym. Sci.,* 26:2693 (1981).

3

Degradation During Blending of ABS/PC and Its Impact Strength Variation

M. S. LEE,* H. C. KAO,* C. C. CHIANG* and D. T. SU*

INTRODUCTION

POLYMER BLENDING PROVIDES an easy method for producing new polymeric materials possessing advantages of its components. The blend of polycarbonate (PC) with acrylonitrile-butadiene-styrene copolymer (ABS) is a typical example. PC contributes the improvements of mechanical and thermal properties to the blend, while ABS provides the benefits of economics, processability and more reliable impact resistance [1]. Recently, our study on this blend showed that the notched Izod impact strength of the blend varied significantly with different ABS given by various suppliers, though all of them had almost the same component content. The impact strength of 0.5 to 15 ft-lb/in had been determined for the blends containing 50 wt.% PC. In this study, the measurement of the melt blending torque, the dynamic viscoelasticity analysis, infrared spectroscopy, and scanning electron microscopy (SEM) were used to explore this difference.

EXPERIMENTAL

The polycarbonate used was Lexan-131 of General Electric Co. Two kinds of locally manufactured ABS, designated here as ABSHKJ and ABSTTS, were used. Before blending, polymer pellets were dried for 6 hours; ABS in an air circulated oven at 70°C, and PC in a vacuum oven at 100°C. The blending of ABS/PC was carried out at 220–250°C using a Warner & Pfleiderer ZSK 30 twin screw extruder. Pellets of the blends obtained from ZSK 30 extruder were dried

This paper was presented at the 43rd Annual Conference, Composites Institute, The Society of the Plastics Industry, Inc., February 1–5, 1988.

*Union Chemical Laboratories, Industrial Technology Research Institute, Hsinchu, Taiwan, Republic of China.

again and then injection molded into the test specimens of 3 mm in thickness. The notched Izod impact test was performed at 23 °C according to ASTM D-256. The fracture surface of the testing bar was vacuum coated with gold and examined by scanning electron microscopy.

A Haake Rheocord torque rheometer system 40 was used to study the variation of the blending torque at 250°C and 70 rpm. The torque appeared at the 15th minute after feeding was defined as τ_{15}.

For IR analysis, thin films of blends were made by compression molding with a hot press at 220–250°C. These films were also used for the dynamic visco-elasticity analysis which was carried out by a Rheovibron DDV-II EA from 25 °C to 180°C at a heating rate of 2°C/min, and frequency of 11 Hz. In addition, films of the ABS/PC blends prepared by solution casting using tetrahydrofuran (THF) as the common solvent were also studied with Rheovibron at the same conditions.

RESULTS AND DISCUSSIONS

Two kinds of ABS, ABSHKJ and ABSTTS, were blended separately with PC at different ratios. As shown in Figure 1, the notched Izod impact strength of ABSHKJ/PC blend increased almost linearly with increasing PC content. But

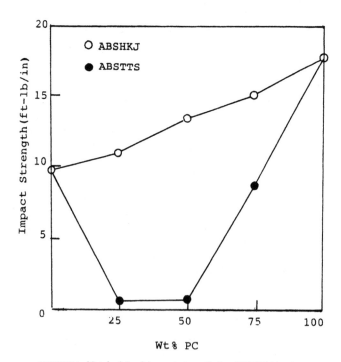

FIGURE 1. Notched Izod impact strength for ABS/PC blends.

FIGURE 2a. SEM micrograph of impact fracture surface of 50-ABSHKJ/50PC blend (×5000).

FIGURE 2b. SEM micrograph of impact fracture surface of 50-ABSTTS/50PC blend (×5000).

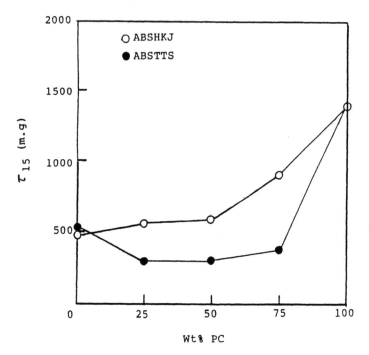

FIGURE 3. Change of τ_{15} with content of PC for different ABS/PC blends.

that of ABSTTS/PC blend dropped to a value of 0.5 ft-lb/in from neat ABS (8.8 ft-lb/in) to the blend containing 25 wt. % PC and rose until PC content reached 50 wt. %. The detriment of the impact strength of ABSTTS/PC was quite severe. The SEM micrographs as shown in Figures 2a and 2b, indicated that a shear-flow happened on the fracture surface of 50 ABSHKJ/50 PC blend, while many pin-holes existed in ABSTTS/PC blend and brittle rupture occurred (Figure 2b).

In this study, ABS/PC were also melt blended in a Haake's torque rheometer. Figure 3 shows that the stable torque, τ_{15}, of ABSTTS/PC was apparently lower than that of ABSHKJ/PC, even lower than that of neat ABS. By comparing Figures 1 and 3, a correspondence was assumed to exist between these two figures.

The compatibility of ABS and PC was studied by the dynamic viscoelasticity analysis. The glass transition temperatures, T_g , of PC and SAN phases in the blends were taken as the peaks of tan δ and were plotted versus PC content in Figures 4a and 4b, respectively. For ABSHKJ/PC blends, T_g of PC and SAN phases shifted closer to each other; at 50/50 composition, 11 °C lower for PC phase in blend than that of neat PC and 5°C higher for SAN phase than that in pure ABS. The shift of T_g implied that ABS/PC was probably partially compatible due to the interaction between PC and SAN phases. However, the change of T_g was somewhat different for ABSTTS/PC blends. T_g of PC phase decreased

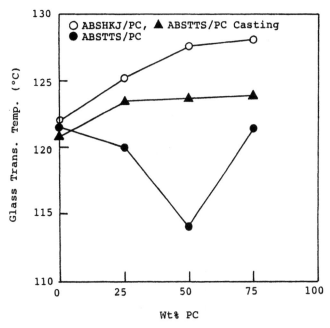

FIGURE 4a. Glass transition temperature of SAN phase in ABS/PC blends.

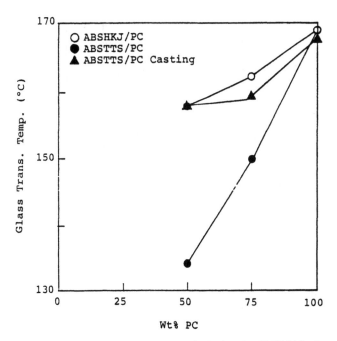

FIGURE 4b. Glass transition temperature of PC phase in ABS/PC blends.

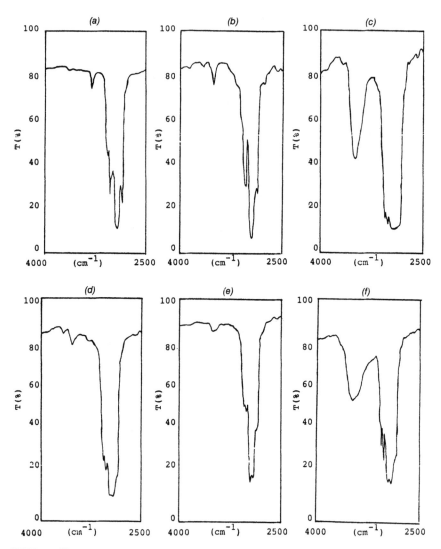

FIGURE 5. IR spectrum: (a) ABSTTS; (b) PC; (c) 50ABSTTS/50PC blend; (d) 50ABSHKJ/50PC blend; (e) 50ABSTTS/50PC blend with stabilizer; (f) 50ABSHKJ/50PC blend with coagulant.

Table 1. The metal content in ABS resins.

	Metal Content (wt%)		
Designation	K	Ca	Mg
ABSHKJ	0.015	—	—
ABSTTS	—	0.029	0.049

more significantly, 34°C lower at 50/50 composition comparing with pure PC. Furthermore, T_g of SAN phase decreased first, 7°C depression at 50 ABS/50 PC comparing 5°C elevation in ABSHKJ/PC blend, then increased with increasing PC content. On the other hand, T_g of PC and SAN phases of the casting films of ABSTTS/PC blends exhibited inner-shift to some extent, which is similar to what happened in the melt blends of ABSHKJ/PC.

It is known that PC in the molten state is easy to degrade when metal salts exist [2]. However, some kinds of metal salts, such as $CaCl_2$ and $MgCl_2$, are usually used as the coagulants in the manufacturing process of ABS [3]. The metal residues of two ABS used in this study are shown in Table 1. It is apparent that the metal salts content of ABSTTS is much higher than that of ABSHKJ, especially the content of calcium and magnesium.

Based on the results mentioned above, we propose that it seems a thermal degradation, induced or promoted by the metal residues in ABS, occurred during the melt blending of ABSTTS/PC. The degradation resulted in the detriment of the impact strength of the blends. The degradation products such as gases and some low molecular weight materials reduced the blending torque and depressed the T_g of PC and SAN phases in the blends due to their plasticizing effect. The pinholes shown in the SEM micrograph of Figure 2b might be caused by the gases existing in the test bar.

The statement of the occurrence of degradation was further proved by the addition of coagulants to the blends of ABSHKJ/PC and/or by the addition of selected stabilizer to the blends of ABSTTS/PC. The results of IR analysis are shown in Figures 5(a–f). A strong peak appearing at 3500 cm^{-1} is observed for ABSTTS/PC blends, but it is very small for pure ABS, PC, and ABSHKJ/PC

Table 2. τ_{15} of ABS, PC, and 50ABSHKJ/50PC blend with coagulant.

Additives	τ_{15}^*, PC	τ_{15}, ABSHKJ	τ_{15}, 50ABSHKJ/50PC
Virgin	1393	488	598
Coagulant (0.1%)**	775	478	239
Coagulant (0.1%)** and stabilizer	1209	478	565

*τ_{15} in m·g.
**Wt percent based on PC.

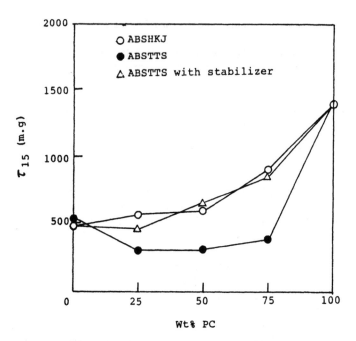

FIGURE 6. The effect of stabilizer on τ_{15} of ABSTTS/PC blends.

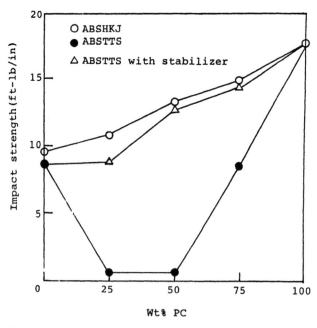

FIGURE 7. The effect of stabilizer on notched Izod impact strength of ABS ABSTTS/PC blends.

blends. This peak diminishes after the addition of stabilizer into ABSTTS/PC blends [Figure 5(e)], and it becomes stronger when the coagulant is added into ABSHKJ/PC blend [Figure 5(f)]. In addition, Table 2 shows the effect of coagulant and stabilizer on the melt mixing torque τ_{15}, of PC, ABSHKJ, and 50 ABSHKJ/50 blend. τ_{15} of ABSHKJ was little changed, but that of PC or a blend dropped severely during the addition of coagulant and maintained at the similar level of the virgin materials during the addition of coagulant and stabilizer at the same time. These results imply that the degradation of ABSTTS/ during melt blending is mainly caused by the induction or promotion effect of the coagulant, existed in ABSTTS as residues, on the degradation of PC. When the stabilizer was added into ABSTTS/PC blends, τ_{15} and the impact strength rose to a level similar to that of ABSHKJ/PC blends as shown in Figures 6 and 7.

CONCLUSION

The notched Izod impact strength of ABS/PC blends differed significantly as the different ABS were used. The blend containing ABSTTS exhibited worse impact strength than that of ABSHKJ/PC due to the severe thermal degradation during melt blending, which occurred preferably in PC phase and was induced by metal salts in ABSTTS. After the addition of a selected stabilizer in 0.1 phr in ABSTTS/PC, the impact strength was greatly improved, and the other properties, such as blending torque and glass transition temperature, could be also brought to the same level of ABSHKJ/PC blends.

REFERENCES

1. Jalbert, R. L. *Mod. Plastics Encyc.*, 61(10A):102 (1984).
2. Gallez, F., R. Legras and J. P. Mercier. *Polym. Eng. Sci.*, 16:276 (1976).
3. Lebovits, A. "Acrylonitrile-Butadiene-Styrene Copolymer," in *Encyclopedia of Polymer Science and Technology, Vol. 1*. 1st edition. New York:John Wiley & Sons, Inc., p. 436 (1965).

BIOGRAPHIES

Mao-Song Lee

Dr. Mao-Song Lee is Manager of polymer processing laboratory of Union Chemical Laboratories (UCL), ITRI, Taiwan, Republic of China. He received his BS and MS degree from National Cheng Kung University in 1970 and 1972, respectively, and Ph.D. in Ch.E. from National Tsing Hua University in 1987. He joined UCL in 1973 and worked in the polymer synthesis laboratory. He has been involved in the research of polymer processing from 1982.

Hsin-Ching Kao

Hsin-Ching Kao is currently an associate researcher at UCL, ITRI, Hsinchu, Taiwan, Republic of China. He received his BS and MS in Ch.E. from Feng-Chia University, Taichung, in 1981 and 1983, respectively. He has been involved in the development and research of polyblends from 1983.

Chih-Cheng Chiang

Chih-Cheng Chiang is currently an associate researcher at UCL, ITRI, Hsinchu, Taiwan, Republic of China. He received his BS in Ch.E. from Chung-Yuan University, Chung-Li in 1980. He has been involved in the development and research of composite materials from 1980.

Der-Tarng Su

Der-Tarng Su is currently an associate researcher at UCL, ITRI, Hsinchu, Taiwan, Republic of China. He received his BS in Ch.E. from National Central University, Chung-Li in 1978 and MS in Polymer Science from National Tsing-Hua University, Hsin Chu, in 1985. He joined UCL and worked mainly in plastics compounding since 1981.

4

Blends of Styrene-Maleic Anhydride Copolymer with ABS

J. J. CHEN,* W. S. LIN,* F. L. LIN* and T. S. TONG*

ABSTRACT: The melt blends of random copolymer of styrene-maleic anhydride (SMA) and commercial acrylonitrile-butadiene-styrene (ABS) resin were prepared by using the twin screw extruder. These blends combine the heat resistance and stiffness of SMA with the toughness and easy processibility of ABS resin. The compatibility of SMA/ABS was investigated by thermal and dynamic mechanical measurements. Inner shifts in transition temperatures (Rheovibron) and a single T_g (DSC) of intermediate value between those of the individual constituents, indicated the partial compatibility of the blends. The morphology of the blends by transmission electron microscopy has proved the good dispersion of the rubber phase. The processibility of these blends evaluated with capillary rheometer exhibited their flow behavior between those of the individual components. In addition, physical properties such as heat distortion temperatures, impact strength, tensile strength and modulus are also discussed in this paper.

INTRODUCTION

THE RANDOM COPOLYMERS of styrene-maleic anhydride (SMA) are distinguished by higher heat resistance, proportional to the MA content [1], than the parent styrenic and ABS families. The MA moiety also yields improved adhesion to glass fiber reinforcement systems. However, its major disadvantage is low impact resistance. ABS resins possess good impact strength, easy moldability and low cost, but low heat distortion temperature. The melt blends of SMA/ABS were prepared by using the twin screw extruder. These blends combine the heat resistance and stiffness of SMA with the toughness and easy processibility of ABS

This paper was presented at the 43rd Annual Conference, Composites Institute, The Society of the Plastics Industry, Inc., February 1–5, 1988.

*Union Chemical Lab. Industrial Technology Research Institute, 321 Kuang Fu Rd. Sec. 2, Hsinchu, Taiwan, Republic of China.

Table 1. Physical and thermal properties of SMA/ABS blends.

Sample Properties	SMA-TM	SMA/ABS (45/55)	SMA/ABS (15/85)	ABS	ASTM Method
Notched Izod Impact (kg-cm/cm, 1/8")	1.31	15.0	20.3	31.4	D-256 Method A
Tensile Strength (@ Break, kg/cm²)	556	474	423	380	D-638
Elongation (%)	3.40	11.0	14.6	15.6	D-638
Tensile Modulus (10^5 kg/cm²)	2.29	1.73	1.49	1.31	D-638
HDT (°C, @ 264 psi, 1/4")	127 (129)*	107 (116)*	98 (109)*	97 (108)*	D-648
Melt Index (g/10 min)	0.82 (230°C/5000g)	0.39 (200°C/5000g)	0.91 (200°C/5000g)	1.3 (200°C/5000g)	D-1238
T_g (quenched, °C, DSC @ N_2)	146	116	107	102	—
T_d (°C, TGA, @ N_2)	377	393	402	418	—

*The results obtained after annealing samples at 80°C for 6 hours.

resin. Several desirable properties, such as glass transition temperature (T_g), heat distortion temperature (HDT) along with tensile strength, of these blends increase with increasing SMA content; whereas, the impact strength and elongation improve with an increase of the ABS moiety.

EXPERIMENTAL

The random copolymer of styrene-maleic anhydride (22% MA) synthesized by solution polymerization was physically mixed with a commercial ABS resin (22% AN) and stabilizers. The above mixtures were pre-dried, then melt compounded in a twin screw extruder. Thermal properties, T_g, T_d (decomposition temperature) were recorded using DSC and TGA. The blend pellets were molded into thin film (0.01 cm in thickness) and test specimens by compression molding and injection molding respectively. Physical properties including HDT, tensile strength, elongation and impact strength were evaluated according to standard ASTM methods. Dynamic mechanical studies were performed with Rheovibron. The processibility of these blends was also evaluated with capillary rheometer.

RESULTS AND DISCUSSION

Physical and Thermal Properties

Physical properties of these blends and individual components are summarized in Table 1. The HDT and tensile strength increase with increasing the SMA

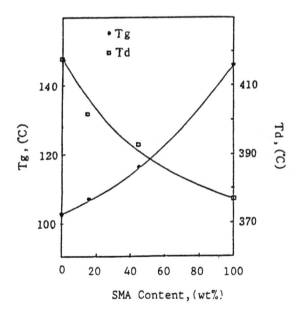

FIGURE 1. Thermal properties as the function of the SMA content.

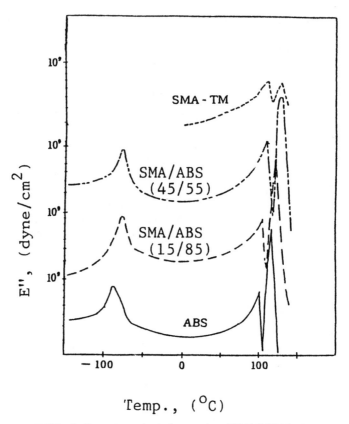

FIGURE 2. Dynamic mechanical properties of SMA/ABS blends.

content; whereas, the impact strength and elongation are enhanced with an increase of the ABS component. DSC thermograms of SMA/ABS (45/55); SMA/ABS (15/85) from room temperature to $160\,^\circ C$ show a single T_g of intermediate value between those of the individual constituents, indicating the partial compatibility of the blends. Additionally, T_g values of these blends are in good agreement with estimated values by the Fox equation [2]. Both T_g and T_d are as a function of the SMA constituent (Figure 1). T_d values of these blends decrease with increasing the SMA constituent, presumably because SMA easily decomposes at high temperature, producing CO_2, H_2O, styrene, ethyl benzene and toluene, etc. [3].

Compatibility

Dynamic mechanical analysis gives the plot of loss modulus, E'', versus temperature for SMA/ABS shown in Figure 2. The test results are numerically

Table 2. Dynamic mechanical properties.

Sample	Transition	Temperature	(°C)
ABS	−85	101	115
SMA/ABS (15/85)	−77	103	121
SMA/ABS (45/55)	−77	111	133
SMA-TM	—	112	139

expressed in Table 2. Inner shifts in transition temperature also indicate the partial compatibility of these blends.

The fracture surface of Izod impact specimens was stained by osmium tetraoxide, followed by transmission electron microscopy observation (Figure 3). Comparison of TEM photomicrographs indicates that rubber phase (dark) is better dispersed in SMA/ABS (44/55) blend than in ABS.

Processibility

Plots of viscosities versus shear rates for these blends and components are shown in Figure 4. Viscosity values of these blends are intermediate between the

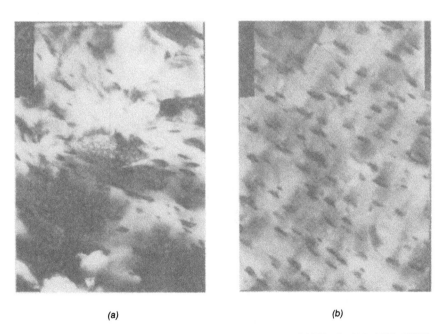

(a)   (b)

FIGURE 3. Transmission electron photomicrograph: (a) ABS, ×20,000; (b) SMA/ABS (45/55), × 15,000.

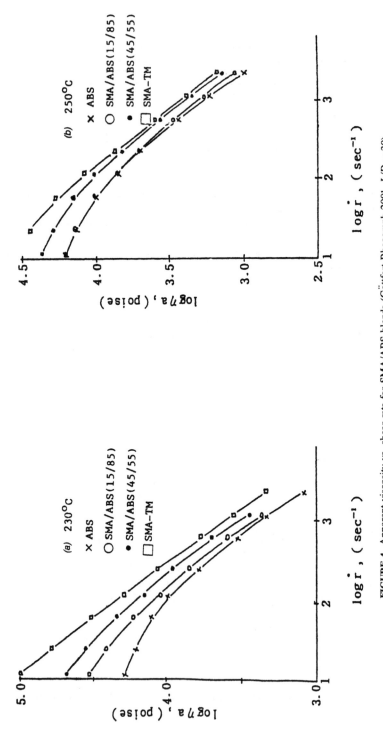

FIGURE 4. Apparent viscosity vs. shear rate for SMA/ABS blends (Göttfert Rheograph 2001, *L/D* =30).

values of ABS and SMA. At higher shear rates and/or higher temperature, the viscosity curves become closer, indicating close processibility for injection molding.

CONCLUSIONS

The SMA/ABS blends provide the heat resistance and stiffness of SMA with the toughness and easy processibility of ABS resin. Partial compatibility of these blends is evidenced in both thermal (DSC) and dynamic mechanical analysis. At higher shear rates and/or higher temperature, the viscosity curves of these blends are close to those of ABS and SMA resins.

REFERENCES

1. Moore, E. R. *Ind. Eng. Chem. Prod. Res. Dev.*, 25:315 (1986).
2. Fox, T. G. Bull. *Am. Phys. Soc.*, 2(2):123 (1956).
3. Orushizaki, M. et al. *Kobunshi Ronbunshu*, 32(6):342 (1975).

5
====

Continuous Graft-Copolymerization Process Using a Two-Stage Type Twin Screw Extruder

TADAMOTO SAKAI*

ABSTRACT: The combination of the twin screw extruder and single screw extruder has been used for continuous reactive processing equipment to control each reaction step, such as the initiation, propagation, disproportionation, and termination reaction. The molten phase graft-copolymerization to a commercial polymer was investigated by way of the ethylene-vinylacetate copolymer—maleic anhydride: styrene monomer.

As a result, it was found that the two-stage type reactive extrusion system is useful as continuous grafting copolymerization equipment, and the modification process which could be rationalized from polymerization to pelletization when compared with the conventional solution polymerization method.

Taking into consideration the reaction temperature and other operation factors, it is possible to control the grafting reaction conversion and production rate when the concentration of monomers and initiator are optimized.

Maleic anhydride and styrene monomer were grafted to the base polymer in an alternating structure.

INTRODUCTION

REACTIVE PROCESSING USING twin screw extruders is becoming very popular in the plastics industry [1–4].

Continuous molten phase graft-copolymerization reaction using high-speed type, two-stage, twin screw extruder has been investigated to modify polymers which are produced by the conventional polymerization process. In this report, ethylene-vinylacetate copolymer and maleic anhydride-styrene monomer grafting system will be discussed. Two purposes of this study are to rationalize manufacturing processes from polymerization to pelletization and to produce new

* Machinery and Electronics Research Center, The Japan Steel Works, Funakoshi-minami 1-6 Aki-ku, Hiroshima, Japan 736.

polymers with improved physical properties. The study has been carried out with respect to graft-copolymerization conversion ratio and output rate versus reaction temperature, monomer and initiator concentration, operational parameters, etc.

MATERIALS

Ethylene-Vinylacetate Copolymer (EVA)

EVA produced by Mitsui-DuPont Petrochemicals Co., Melt Index (MI) 165 and 16, VA content 30%.

Monomer

- styrene monomer (ST) MW = 108, commercial first grade
- maleic anhydride (MAA) commercial first grade
- benzoil peroxide (BPO) commercial first grade
- tertial-butyl peroxiacetate (BPA) commercial grade
- di-tertial-butyl peroxide (DBP) commercial grade

EXPERIMENTAL DEVICE AND METHOD

Experimental Device

Two-stage type twin screw extruder is shown in Figure 1. This extruder is the combination of a twin screw extruder unit and a single screw extruder unit. The twin screw unit is composed of non-intermeshing counter-rotating twin screws which are used for solid conveying, melting, mixing and kneading. This unit was used mainly for the initiation reactor for the copolymerization process. The

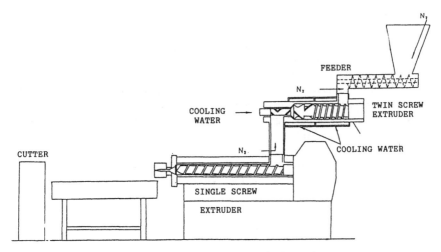

FIGURE 1. Continuous graft copolymerization equipment.

single screw unit was used for propagating and terminating the reaction and for discharging and pelletizing of molten polymer through the dies.

The main specifications of the extruder are as follows:

1. Twin screw extruder unit
   - *screw geometry:* full-flight and rotor as shown in Figure 2
   - *screw diameter:* 120 mm, screw speed: 150–900 rpm; two screws rotate with the speed ratio of 1:1.18
   - *throttle gap:* 2–30 mm, adjustable by moving the twin screw cylinder unit forward or backward

   Material feeder capacity: 50–1000 kg/h

   Cylinder and screw heating/cooling: steam/water
2. Single screw extruder unit
   - *screw diameter:* 120 mm, length: 12 L/D
   - *cylinder heating:* aluminum casting heaters
   - *screw geometry:* full-flight screw; channel depth: 20 mm

Experimental Method

Raw materials, EVA, MAA, ST and peroxide were pre-blended with each other using a Henschel type mixer prior to feeding to the extruder. Initiators were mixed with ST monomer, and ST monomer penetrated easily into the EVA polymer.

Polymer temperature (reaction temperature) at the twin screw extruder unit was controlled mechanically through the adjustment of screw speed and throttle gap under adiabatic operational conditions. The relationship between polymer temperature and screw speed is shown in Figure 3.

Analytical Method

1. *Grafting conversion analysis*—Grafted extrudates were purified using xylene solvent and methylalcohol. After having removed unreacted monomer, samples were vacuum-dried. Graft conversion was defined as (MAA content of purified polymer/MAA initially added to the polymer).

   MAA content of purified polymers was determined by the (1/10)N alcoholic KOH titration method after dissolving benzene-isopropyl alcohol (7:3). The indicator used here was phenolphthalein. In the calculation, MAA was considered as a monobasic acid. No correction has been made, even if some anhydride changed dibasic acid and the measured values showed more than 100%.

2. *Melt Index value measurement*—MI values were measured at 190°C and 2160 g load. In the case of very low melt viscosity EVA, a 325 g load was used, and MI values were calculated from the following equation:

$$MI_{2160} = 10^{1.39} \times FR^{0.921}$$

where FR is the flow rate of molten EVA under a 325 g load (g/10 min).

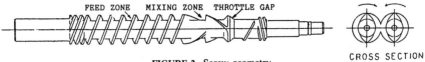

FEED ZONE   MIXING ZONE   THROTTLE GAP

CROSS SECTION

FIGURE 2. Screw geometry.

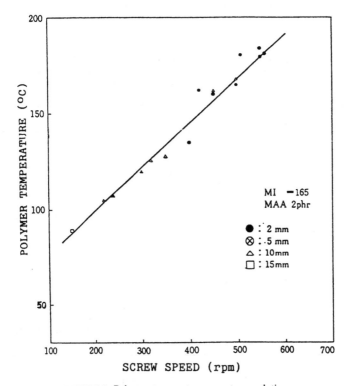

FIGURE 3. Polymer temperature vs. rotor revolution.

Table 1. Graft copolymerization in absence of initiating agent.

			Twin Screw Extruder					
Polymer MI	MAA phr	ST phr	Temp. °C	Speed rpm	Gap mm	Tp °C	Product MI	Graft %
16	2.0	2.125	192	500	5	204	13.4	73
16	2.0	2.125	100	150	20	155	20	13

Table 2. Influence of initiators (base polymer MI = 165).

Initiator				Twin Screw Extruder					
Peroxide	mole/kg	MAA phr	ST phr	Temp. °C	Speed rpm	Gap mm	Tp °C	Product MI	Graft %
BPO	2.15×10^{-3}	2.0	2.125	128	320	10	134	57	145
	4.1×10^{-3}	2.0	2.125	126	300	10	129	47	137
BPA	7.1×10^{-3}	2.0	2.125	181	700	5	169	69	136
	3.1×10^{-3}	2.0	2.125	180	510	5	169	57	128
	1.21×10^{-3}	2.0	2.125	181	560	5	174	34	126
DBP	5.3×10^{-3}	2.0	2.125	165	500	2	186	62	132

Table 3. Influence of initiators (base polymer MI = 16).

Initiator				Twin Screw Extruder					
Peroxide	mole/kg	MAA phr	ST phr	Temp. °C	Speed rpm	Gap mm	Tp °C	Product MI	Graft %
BPA	2.7×10^{-3}	2.0	2.125	158	250	10	176	4.0	130
BPA	1.4×10^{-3}	2.0	2.125	152	250	10	173	5.7	134
DBP	2.7×10^{-3}	2.0	2.125	188	330	5	178	5.4	125

EXPERIMENTAL RESULTS

Graft-Copolymerization without Peroxide

It is well-known that mechanical shearing action generates polymer radicals due to main chain scission. Table 1 shows experimental results of continuous grafting polymerization without any kind of initiator. EVA used here was a higher molecular weight polymer in order to generate more radicals under high screw speed. The final grafting conversion obtained after going through the second extruder unit was at most 73%. The graft conversion at the exit of the twin screw unit was only 2.0%. These experiments showed that radical generation was so low that a catalyst should be used to accelerate the grafting reaction.

Graft-Polymerization with Peroxide

Tables 2 and 3 show the experimental results of continuous graft copolymerization using peroxide as initiation agents. Decomposition temperature of the peroxides becomes higher in order of BPO < BPA < DBP. When peroxides were used, grafting polymerization was completed easily. Even if the concentration of the initiator was as low as BPA 7.6×10^{-4} mol/kg, the completion of the grafting reaction was possible. However, there was a tendency for the residual monomer concentration to increase slightly in proportion to the decrease of the initiator concentration. The optimum concentration of the initiator should be decided with respect to the polymer *MI* obtained and final graft conversion. Too high a concentration of initiator causes gel formation when the polymer is dissolved in benzene.

Table 4 shows the influence of polymer temperature on grafting conversion of the EVA-MAA-ST system. A close relationship exists between polymer temperature and graft conversion obtained from each extrusion unit, especially twin screw extruder unit. This means that the continuous graft copolymerization

Table 4. Influence of twin screw operating conditions (MI = 165).

| Initiator | | MAA | ST | Twin Screw Extruder | | | Tp | Product | Graft |
| | | | | Temp. | Speed | Gap | | | |
Peroxide	mole/kg	phr	phr	°C	rpm	mm	°C	MI	%
BPO	4.1×10^{-3}	2.0	2.125	89	150	15	133	51	23* 133
BPO	4.1×10^{-3}	2.0	2.125	105	220	10	132	52	66* 136
BPO	4.1×10^{-3}	2.0	2.125	126	320	10	134	47	121* 145
BPA	5.3×10^{-3}	2.0	2.125	119	300	10	154	60	60* 135
BPA	5.3×10^{-3}	2.0	2.125	137	450	10	161	58	85* 138
BPA	5.3×10^{-3}	2.0	2.125	160	450	5	166	70	110* 130
DBP	5.3×10^{-3}	2.0	2.125	135	400	5	175	59	29* 136
DBP	5.3×10^{-3}	2.0	2.125	165	500	2	186	62	109* 132

*Grafting conversion of polymer extrudated at twin screw unit.

Table 5. Influence of monomer concentration (MI = 165).

Initiator		MAA	ST	\multicolumn{5}{c}{Twin Screw Extruder}				
Peroxide	mole/kg	phr	phr	Temp. °C	Speed rpm	Gap mm	Tp °C	Product MI	Graft %
BPA	3.1×10^{-3}	2.0	2.125	180	510	5	169	57	128
BPA	6.1×10^{-3}	4.0	4.250	181	570	5	175	43	113
BPA	1.2×10^{-2}	8.0	8.500	181	370	5	179	18	108

process can be controlled under good mixing conditions. Optimum polymer temperature should be determined from consideration of monomer boiling point, peroxide decomposition characteristics, polymer degradation, etc.

The role of the two-stage type reaction extruder is classified in one of the following ways:

(a) The first stage twin screw extruder unit is used for dispersing raw materials, such as polymer, monomer and initiator. In this case, polymerizing mixture should be kneaded and extruded at as low a temperature as possible. Grafting reaction is almost completed at the second stage extruder with long residence time.

(b) Half of the grafting reaction should be completed at the first twin screw unit under strong mixing and kneading conditions. The second stage extruder is used for the completion of the polymerization reaction, devolatilization of residual monomers [5] and shaping of products.

(c) Grafting reaction is almost completed at the first stage twin screw extruder. As a result, this way is a single stage reactor system. In this case, the twin screw extruder with long L/D should be used.

The proper use of a two-stage type reactor system is described by case (b). The advantages of the two-stage type extruder system is the independent control of temperature, residence time and mixing at each extrusion reaction stage, such as initiation, propagation and termination reaction. Specifically, a high-speed type twin screw extruder has the capability to control precise polymer temperature and mixing conditions for the mixture of a high viscosity polymer and a low molecular weight monomer through strong viscous heat dissipation adjustment.

Table 6. Influence of monomer concentration (MI = 16).

Initiator		MAA	ST	\multicolumn{5}{c}{Twin Screw Extruder}				
Peroxide	mole/kg	phr	phr	Temp. °C	Speed rpm	Gap mm	Tp °C	Product MI	Graft %
BPA	3.1×10^{-3}	2.0	2.125	181	280	5	186	9.2	116
BPA	6.1×10^{-3}	4.0	4.250	182	250	5	187	6.5	110

The influence of monomer concentration on graft-copolymerization is shown in Tables 5 and 6. These data show that higher MAA grafting ratio was obtained at the range of 2.0–8.0 phr. There was not a large difference between a higher *MI* polymer and a lower one, if polymer temperature was kept within an adequate extrusion range.

DISCUSSION

Copolymerization Reaction Rate

To investigate the mechanism of the grafting reaction in the twin screw extruder, reaction kinetics should be discussed.

Polymerization reaction rate is generally expressed by the following equation.

$$Kp = A \exp(-E/RT)$$

where

E = activation energy
T = absolute temperature
R = gas content
A = constant

If it is assumed that the polymerization rate Kp is expressed as $[(M)_t/(M)_{initial}]$ and the equation shown above is reasonable, there is a linear relationship between log Kp and $1/T$. Constant A and activation energy E can be obtained from this line. $(M)_t$ is the grafted MAA-ST content (phr) measured after purification and $(M)_{initial}$ is the initial concentration of MAA-ST. This relation is shown in Figure 4. The average residence time of this reactor system is 1.5 min (100 kg/h output rate) for the twin screw extruder unit and 7.0 min for the single extruder unit. In Figure 4, plotted points were corrected under the assumption that the measured graft conversion of 120% corresponded to the true maximum conversion of 100%.

From these data, MAA-ST grafting conversion rates are obtained as follows:

1. Without peroxide:

$$Kp = 3.86 \times 10^3 \exp(-5.8 \times 10^3/RT)$$

2. BPO 4.1 $\times$ 10^{-3} mol/kg:

$$Kp = 1.00 \times 10^{11} \exp(-16.4 \times 10^3/RT)$$

3. BPA 5.3 $\times$ 10^{-3} mol/kg:

$$Kp = 1.34 \times 10^5 \exp(-6.4 \times 10^3/RT)$$

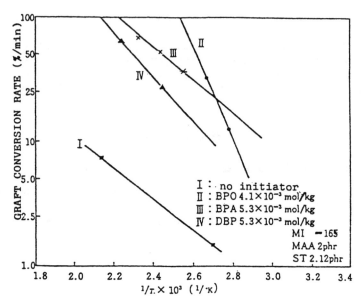

FIGURE 4. Rates of graft polymerization.

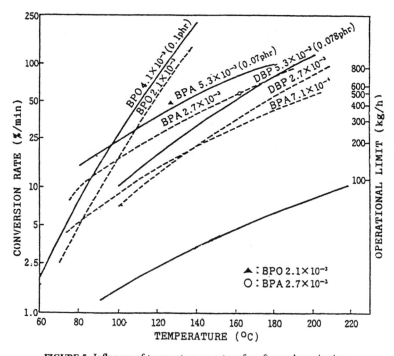

FIGURE 5. Influence of temperature on rates of graft copolymerization.

4. DBP 5.3 × 10⁻³ mole/kg:

$$Kp = 1.15 \times 10^6 \exp(-8.65 \times 10^3/RT)$$

Figure 5 shows the relation between polymer temperature and grafting conversion rate from these reaction rates. Using these data, we can calculate the expected grafting capacity of this continuous molten state polymerization reaction system. For example, the polymerization capacity to produce EVA-MAA-ST graft polymer will be only 28 kg/h in the case of no initiator use. However, the capacity jumps to 260 kg/h if DBP is used as the initiator (470 kg/h, if BPA and 1900 kg/h, if BPO), at polymer temperature 140°C. Dashed lines in Figure 5 were calculated from the following equation under the assumption that the grafting polymerization rate was increased in proportion to [initiator concentration]$^{0.5}$ the same as a general radical polymerization.

$$Kp = B[K]^{0.5} \exp(-E/RT)$$

where

B = constant
$[K]$ = initiator concentration

In the case of BPA, measured value corresponded well with the calculated dashed line in Figure 5.

Branching Structure

MI values of grafted polymers obtained were much lower than the base EVA polymer, mainly because of the cross-linkage reaction due to peroxide radicals, and partly because of a grafted component, such as branching styrene. It is important to prevent this reduction of *MI* value of the polymer in order to maintain good fluidity. Gel formation in the polymer and drastic *MI* drop have been observed when a polymer is reacted with monomers and peroxide under poor mixing conditions in an inadequate kneader, such as a single screw extruder. Grafting reaction of molten polymer is dominated by the diffusion of monomer and peroxide into the polymer. Therefore, it is necessary to disperse additives homogeneously at the melting and mixing zone of the reactor. This is the reason why we utilize a twin screw extruder unit as the initiation reactor.

Figure 6 shows the relationship between MI changes and initiator concentration. And Figure 7 shows the relation between *MI* values of extrudates and grafting conversion rate. There was a tendency that higher initiator concentration and conversion rate brought about larger *MI* drop under the same mixing condition.

From Figure 6, the relationship between the initiator concentration and polymer MI obtained was as follows:

MAA-ST 2.0 phr: $[MI] = 72 \exp(-30[K])$

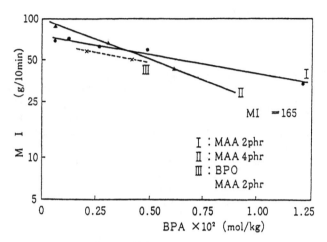

FIGURE 6. Influence of initiator concentration on *MI* values.

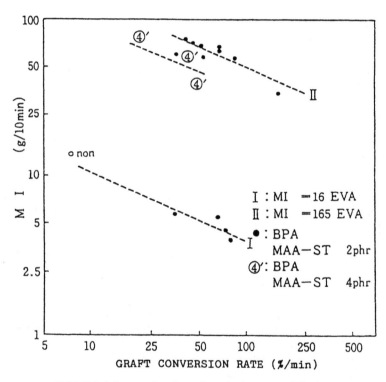

FIGURE 7. Influence of graft copolymerization rate on *MI* values.

MAA-ST 4.0 phr: [MI] $= 100 \exp(-60[K])$

Larger graft quantity caused larger *MI* drop. Generally a faster polymerization rate leads to a shorter molecular weight polymer chain in the usual radical polymerization. From Figures 6 and 7, it was estimated that shorter but larger numbers of graft branches brought about larger *MI* drop, if the cross-linkage degree were assumed to be the same.

The influence of grafted branching chains and their length should be further studied in detail.

Reaction Temperature Control

Figure 8 shows the thermogram to investigate the heat of reaction using the differential scanning calorimetry. The heat generation of this polymerization obtained from Figure 8 was about 1.3 kcal/kg. This corresponds to the 10–20% energy necessary to raise the polymer temperature 140°C. Therefore, the heat removal from the extruder is not so important, when compared with the bulk polymerization with larger heat reaction.

Graft Polymer Structure

Figure 9 shows the infrared spectrum of purified EVA-MAA-ST graft copolymer obtained in this study. Peaks at 1860 cm⁻¹ and 1780 cm⁻¹ showed that MAA component was grafted to EVA and a peak at 700 cm⁻¹ showed ST component. This sample is 8 phr grafted copolymer. Figure 10 shows the relationship between ST and MAA component absorbance. The 1020 cm⁻¹ peak of EVA was used as the inner standard. Figure 10 means that MAA and ST were combined with constant monomer ratio.

Figure 11 shows the relationship between MAA graft conversion and ST monomer content. The MAA conversion value attained a maximum at the level of (MAA)/(ST) mole ratio 1.0–1.5. This means that MAA and ST formed the

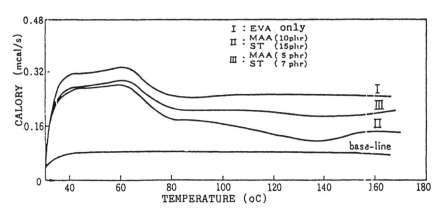

FIGURE 8. DSC thermogram.

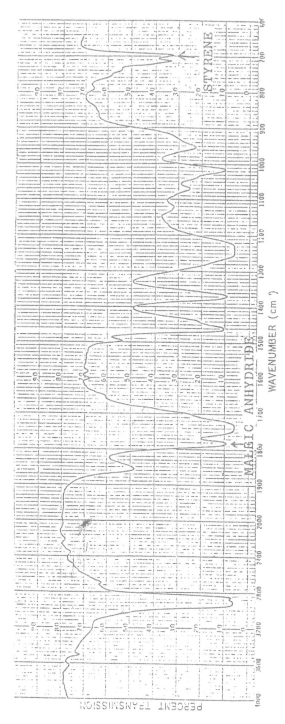

FIGURE 9. Infrared spectra of grafted ethylene-vinylacetate copolymer.

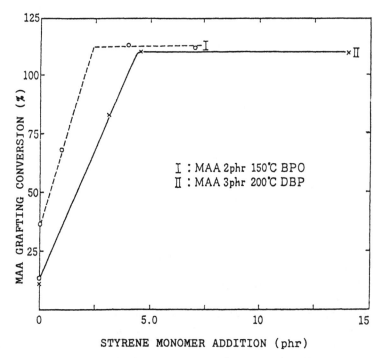

FIGURE 11. Influence of ST concentration on MAA conversion.

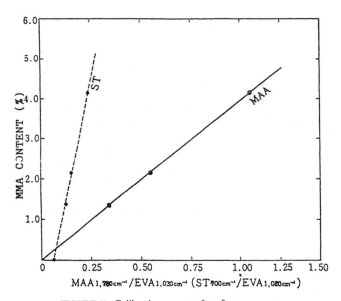

FIGURE 10. Calibration curves of graft components.

Table 7. Rates of copolymerization between various radicals and monomers (unit: min⁻¹).

Monomer \ Radical	VAc	AN	MMA	ST
VAc	2,040	104	29	3.3
AN	3×10^4	425	425	340
MMA	20×10^4	2,350	575	340
ST	20×10^4	1×10^4	1,250	178
MAA	3.7×10^4	71	86	1.8×10^4

VAc = vinyl acetate; AN = acryl nitril; MMA = methyl methacrylate.

alternating copolymer. The reason that the ST concentration was slightly higher than that of MAA was that ST was easily volatilized during extrusion (boiling point: ST, 145°C and MAA, 200°C).

Table 7 shows the reactivity among polymer radicals and monomers [6]. When compared with the reactivity of EVA-MAA and EVA-ST, the latter is larger. Therefore, it seems that the EVA-ST graft copolymerization occurs first, then EVA-ST-MAA alternating copolymerization appears at the second stage. First, copolymerization would occur easily, because ST monomer penetrated into EVA polymer at the pre-blending stage. This is very similar to the radiation graft copolymerization for polyethylene and other monomers in which the monomer easily penetrates into polyethylene and has a tendency to combine readily with the parent polymer [7].

CONCLUSION

Using commercial ethylene-vinylacetate copolymer, continuous graft copolymerization of the molten state has been investigated. The combination of high-speed type, twin screw extruder and single screw extruder was used as the copolymerization reactor system. In this study it was found that the twin screw extruder unit acting as an initiation reactor played an important role in the melting and dispersing of polymeric material, monomers and peroxide which prevented the cross-linking reaction from occurring in the polymer. The single screw extruder unit with longer residence time worked as the propagation and termination reactor. Major operational parameters are polymer temperature, feeding rate (hold-up time), initiator and its concentration. The adhesive properties of grafted ethylene-vinylacetate copolymer obtained in this study were very much improved for aluminum foil, when compared with the untreated EVA.

Recently reactive extrusion using a twin screw extruder has become more important in the plastics industry. The two-stage type twin screw extruder system, such as the combination of twin screw–single screw extruder and/or twin screw–twin screw extruder is increasing as more sophisticated chemical reactors are built.

This study was presented at SPE ANTEC 88 held in Atlanta, in April, 1988.

REFERENCES

1. Tucker, S. and R. Nichols. *Plastics Engineering.* p. 27 (May 1987).
2. Baker, N. et al. *Polymer Engineering and Science,* 27(20):1634 (1987).
3. Lambla, M. et al. *Polymer Engineering and Science,* 27(20):1221 (1987).
4. For example, *Modern Plastics,* p. 28 (July 1987).
5. Sakai, T. and N. Hashimoto. *SPE ANTEC 86.* Boston, p. 860 (1986).
6. Bagdasarian. Theory of Radical Polymerization, Asakura Shoten.
7. Ordin, D. *J. Polymer Science,* 54:511 (1961).

6

Mechanical, Dynamic Mechanical Properties and Morphology of Modified Polyphenylene Oxide Blends

CHEN-CHI M. MA,* CHIEN-MEI HSIAO* and HSI-PAI HSU**

ABSTRACT: Impact resistance of PPO/HIPS blend can be improved by adding styrene-butadiene copolymer (SBS) or methyl methacrylate-butadiene styrene copolymer (MBS) modifier. Mechanical properties of PPO/HIPS/SBS and PPO/HIPS/MBS blends have been studied. It was found that SBS modifier is better than MBS modifier in the enhancement of impact strength and elongation of blends. However, the decreases of tensile, flexural strength and modulus of MBS modifier are less than those of SBS modifier. Heat distortion temperature (HDT) shows no significant change even after 20 phr of either rubber modifiers have been added.

Dynamic mechanical properties (E', E'', and tan δ) were also studied to investigate the compatibility between the modifier and PPO/HIPS matrix. Although PPO/HIPS/RUBBER blends show multiphase it was found that the tan δ peak of rubber domain of the HIPS resin shifted when modifiers were added. Results indicate modifiers show no significant influence on the T_g of PPO/PS matrix. The morphology of the fracture surface of blends was also investigated using scanning electron microscopy (SEM). Results show that the interaction force between SBS and PPO/HIPS is better than that of MBS and PPO/HIPS.

INTRODUCTION

POLY(2,6-DIMETHYL-1,4-PHENYLENE OXIDE) (PPO) shows high distortion temperature, good mechanical properties at elevated temperature, excellent electrical insulation, etc. However, since PPO contains a linear bonded aromatic structure,

---

*Institute of Chemical Engineering, National Tsing Hua University, Hsin-Chu, Taiwan, Republic of China.

**Catalyst Research Center, China Technical Consultants, Inc., Hsin-Chu, Taiwan, Republic of China.

it is stiff and has a relatively high melt viscosity. Blending high impact polystyrene (HIPS) with PPO can reduce the melt viscosity and increase impact strength of polymer matrix slightly [1,2]. Furthermore, an adequate impact modifier is required to improve the impact strength of the blends.

The addition of a dispersed rubbery phase to enhance the toughness of plastics has been recognized and exploited commercially [3]. The mechanical properties of "rubber-modified thermoplastics" depend primarily on the following factors:

(1) Types of rubber and plastics
(2) The ratio of continuous phase (plastics) to dispersed phase (rubber)
(3) The morphology of the rubber particles
(4) The interaction force between rubber phase and plastic phase
(5) The glass transition temperature (T_g) of the rubber added
(6) Processing conditions (temperature, shear rate, etc.) [4]

In this study, styrene-butadiene copolymer (SBS) and methyl methacrylate-butadiene-styrene copolymer (MBS) were selected as impact modifiers. PPO/HIPS/RUBBER was mixed by mechanical blending in molten state. Mechanical properties (impact strength, tensile strength, modulus, elongation) and heat distortion temperature (HDT) have been investigated. Dynamic mechanical properties (E', E'', and tan δ) were studied by utilizing a dual cantilever in the temperature range of −130°C to 135°C. Compatibility between modifier and PPO/HIPS matrix is investigated. Scanning electron microscope was used to study the morphology of PPO/HIPS/RUBBER blends.

EXPERIMENTAL

Materials

Poly(2,6-dimethyl-1,4-phenylene oxide) resin used in this study was synthesized by the Catalyst Research Center, China Technical Consultants, Inc., Taiwan, R.O.C. Intrinsic viscosity (IV) of PPO in chloroform is 0.5 dl/g. The high impact polystyrene (HIPS) used in this research was TAITA 551 (Taita Chemical Co. Ltd., Taiwan, R.O.C.). The styrene-butadiene copolymer (SBS) used was TUFPRENE A (Asahi Chemical Industry Co. Ltd., Japan) which contains 40 wt% styrene. The methyl methacrylate-butadiene-styrene copolymer (MBS) used was BTA-3N2 (Kurha Chemical Industry Co. Ltd., Japan).

Preparation of Blends

Figure 1 shows the flow chart of the experiment in this research. The twin-screw extruder used is a Rheocord system 400, Haake Buchler Co. Ltd., U.S.A. The extruder barrel was set at the following temperature profile: 265°C (zone 1), 270°C (zone 2), 280°C (zone 3), and 275°C (die). Injection molding machine used is a Niigate model SKM, Japan. The barrel was set at the following temperature profile: hopper, 110°C; zone 1, 255°C; zone 2, 255°C; zone 3, 255°C; die, 255°C mold; 90°C.

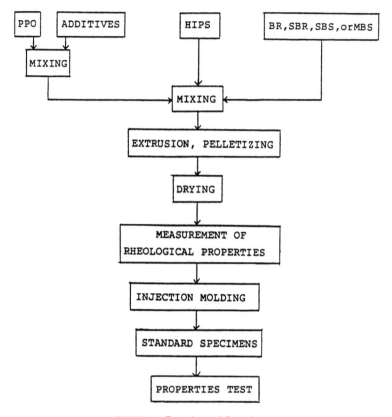

FIGURE 1. Experimental flow chart.

Instruments

MECHANICAL PROPERTIES TESTS

The Izod impact strength of notched sample was measured at room temperature, according to ASTM-D256. Tensile tests were studied by an Instron 1123 type Tester at a crosshead speed of 50 mm/min, according to ASTM-D638 type I. HDT was studied according to ASTM-D648, load pressure is 264 psi, and heating rate is 2°C/min.

DYNAMIC MECHANICAL PROPERTIES STUDY

A Dynamic Mechanical Thermal Analyzer (DMTA) manufactured by the Polymer Laboratories, U.K. was utilized to study the dynamic mechanical properties of polyblends. Test specimens were prepared by compressing the blends to a 0.9 mm film and then cut into a 4 cm long and 0.6 cm wide strip. Data were obtained at 1 Hz while the samples were heated from −130°C to 135°C at a heating rate of 3°C/min.

MORPHOLOGY

The fracture surfaces of impact test were coated with Au and examined with a Camscan Scanning Electron Microscope, U.K.

RESULTS AND DISCUSSION

Mechanical Properties

Figure 2 shows the Izod impact strength of PPO/HIPS (50:50 wt%) with different levels of MBS and SBS modifiers at room temperature. The impact

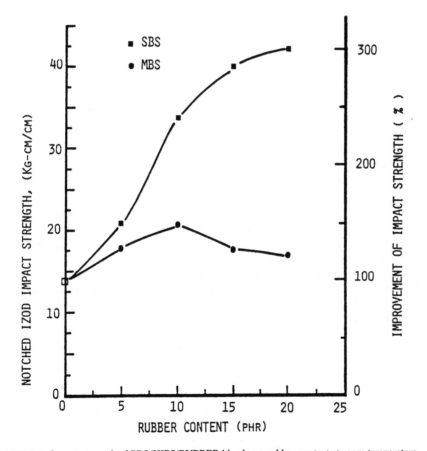

FIGURE 2. Impact strength of PPO/HIPS/RUBBER blends vs. rubber content at room temperature.

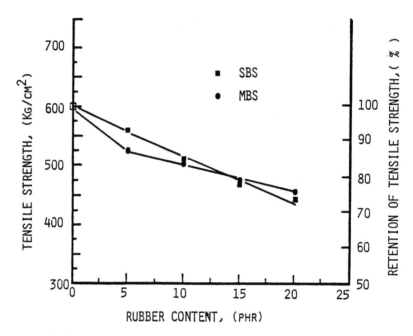

FIGURE 3. Tensile strength of PPO/HIPS/RUBBER blends vs. rubber content at room temperature.

strength of PPO/HIPS/SBS blends increases sharply (from 20.96 kg-cm/cm to 33.75 kg-cm/cm) when adding 5–10 phr SBS copolymer to the PPO/HIPS system. However, impact strength of blends increases only slightly by adding 10 phr MBS modifier, and impact strength decreases gradually at higher modifier level. It indicates that SBS modifier is better than MBS modifier in the enhancement of impact strength of blends. Since SBS modifier is a block copolymer, styrene segments can provide good interfacial force with the PPO/HIPS matrix, and avoid phase separation [5].

Effects of impact modifier contents on the tensile strength, elongation and tensile modulus of PPO/HIPS/SBS and PPO/HIPS/MBS blends are shown in Figures 3, 4, and 5. From Figure 3 one can find that adding 5 phr of either modifiers will reduce the tensile strength of blends due to the flaws occurring in the PPO/HIPS/RUBBER blends. It was shown in Figure 4 that SBS modifier can enhance the elongation of blends more effectively than that of MBS modifier. After adding 20 phr SBS copolymer, the elongation of blends increased to 42.2% from 13.3%. However, the increase of elongation was only 7.8% when adding 20 phr MBS. The tensile modulus of PPO/HIPS/MBS is higher than that of PPO/HIPS/SBS blends in the modifier range of 5–20 phr as shown in Figure 5.

MBS modifier is usually synthesized by grafting methyl methacrylate and styrene onto a styrene-butadiene rubber with a small degree of cross-linking density, while SBS has the characteristics of thermoplastic elastomer. MBS copoly-

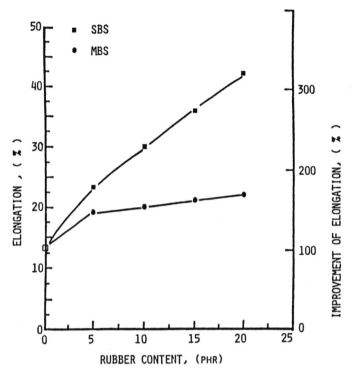

FIGURE 4. Elongation of PPO/HIPS/RUBBER blends vs. rubber content at room temperature.

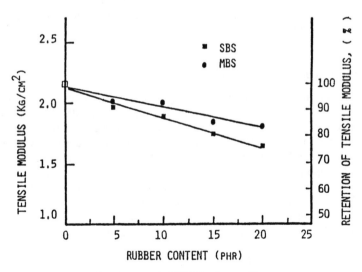

FIGURE 5. Tensile modulus of PPO/HIPS/RUBBER blends vs. rubber content at room temperature.

Table 1. Comparison of HDT of various PPO/HIPS/RUBBER blends.

Composition	HDT (°C)
PPO/HIPS	121
PPO/HIPS/SBS (5)	121
PPO/HIPS/SBS (10)	120
PPO/HIPS/SBS (15)	121
PPO/HIPS/SBS (20)	120
PPO/HIPS/MBS (5)	119
PPO/HIPS/MBS (10)	120
PPO/HIPS/MBS (15)	119
PPO/HIPS/MBS (20)	119

mer is stiffer than SBS copolymer and its tensile modulus is higher than that of SBS. According to the rule of mixtures, MBS modifier can provide higher strength and tensile modulus than that of SBS modifier. Hence, the decrease of tensile strength and modulus of MBS modifier is less than those of SBS modifier. However, due to the poor adhesion between methyl methacrylate and PPO/HIPS matrix, cracks will occur at the boundary phase first, then propagate, and finally a catastrophic rupture occurs. Consequently, adding MBS modifier shows a little improvement in impact strength and elongation in blends.

Table 1 summarizes the heat distortion temperature of PPO/HIPS/RUBBER blends. There is no significant declination in HDT when adding impact modifier, even 20 phr of either SBS or MBS copolymer was added. These data suggest that these blends are still rigid enough to support the load applied even after 20 phr of rubber modifier was added to the blends.

Dynamic Mechanical Properties

The storage modulus (E') and loss tangent (tan δ) of the pure SBS and MBS copolymer are shown in Figures 6 and 7, respectively. For SBS block copolymer, the existence of two loss peaks is evident. The one at higher temperature (91°C) is attributed to the movement of long chain segments of styrene groups, while the relaxation peak at the low temperature (-71°C) is due to the butadiene segments in the SBS copolymer. For MBS copolymer, the low temperature relaxation peak (-44°C) is caused by the styrene-butadiene chain, while the high temperature relaxation peak (88°C) may be due to the grafted methyl methacrylate and styrene components. Since MBS copolymer is stiffer than SBS copolymer, the log E' of MBS is larger than that of SBS in the temperature range of -130°C to 100°C.

Storage modulus (E'), loss modulus (E''), and loss tangent (tan δ) versus temperature at various blend compositions are summarized in Figures 8 to 13.

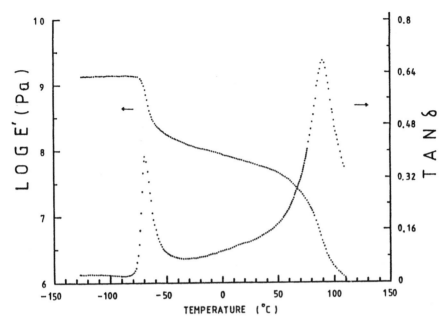

FIGURE 6. Storage modulus (E'), loss tangent (tan δ), vs. temperature for SBS.

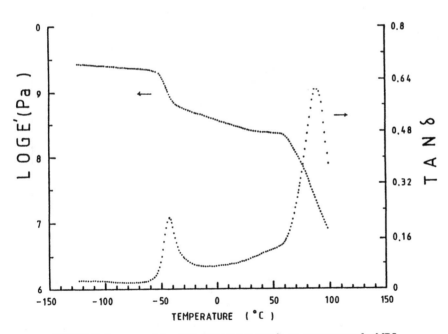

FIGURE 7. Storage modulus (E'), loss tangent (tan δ), vs. temperature for MBS.

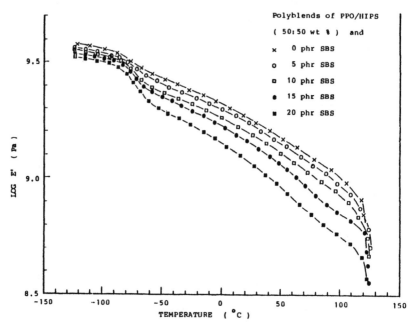

FIGURE 8. Storage modulus (E') vs. temperature for PPO/HIPS/SBS blends at different SBS contents.

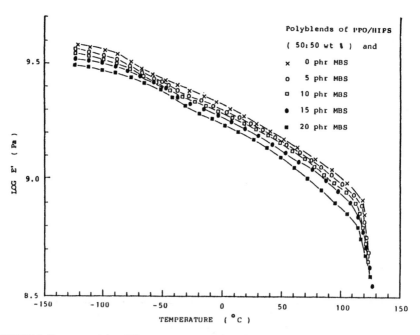

FIGURE 9. Storage modulus (E') vs. temperature for PPO/HIPS/MBS blends at different MBS contents.

66

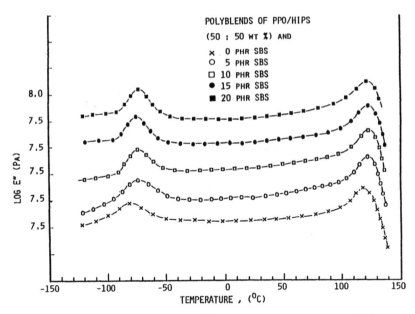

FIGURE 10. Loss modulus (E'') vs. temperature for PPO/HIPS blends at various SBS contents.

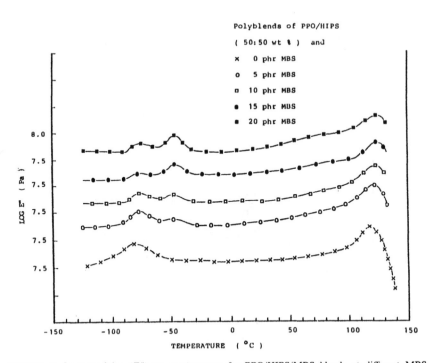

FIGURE 11. Loss modulus (E'') vs. temperature for PPO/HIPS/MBS blends at different MBS contents.

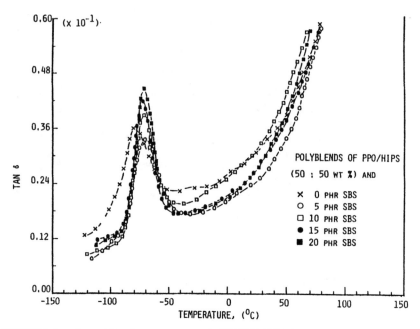

FIGURE 12. Loss tangent (tan δ) vs. temperature for PPO/HIPS/SBS blends at various SBS contents.

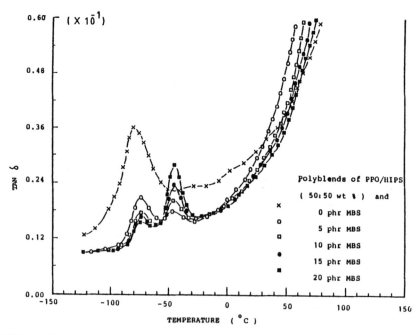

FIGURE 13. Loss tangent (tan δ) vs. temperature for PPO/HIPS/MBS blends at different MBS contents.

68

Figures 8 and 9 show the log E' of blends decrease with the increasing of MBS or SBS modifier content. Below the T_g of rubbery phase, the changes of log E' with various rubber content are small. Above the T_g of rubbery phase the deviations become larger and these phenomena are even predominant for SBS blend systems. For all of the PPO/HIPS/RUBBER blends in the research, the decrease of log E' of blends below 124°C is less than one order of magnitude. These results illustrate that rubbery phase was dispersed in the PPO/HIPS matrix. Above 124°C, log E' of all the blends decreases rapidly.

Figures 10 and 11 show the log E'' of PPO/HIPS/RUBBER blends. The presence of two peaks can be seen in PPO/HIPS (50:50 wt%) blends. The first one occurred at the temperature of 124°C which is attributed to the completely miscible matrix, PPO/HIPS. The other peak occurred at the temperature −81°C which is due to the dispersed, grafted rubbery domain of HIPS that is immiscible with the PPO/HIPS matrix. The high temperature peaks of log E'' are very similar for SBS and MBS modifiers. These blends also show multiphase structure, hence, adding modifier (≤ 20 phr) will not change the T_g of the PPO/HIPS.

When adding SBS modifier, the low temperature peak occurred at −72°C independent of the SBS content. While adding MBS modifier, two low temperature peaks appeared, one at −76°C, the other at −45°C; the locations of peaks are independent of MBS content.

Figure 12 shows the effect of SBS contents on the loss tangent (tan δ) of PPO/HIPS/SBS blends. At high temperature, tan δ is too high to be detected due to the limit of the instrument. At low temperature, the α-relaxation peak of butadiene domain of HIPS rises 9°C after adding SBS rubber and approaches the relaxation peak of butadiene segments in SBS copolymer. Two discrete peaks become one narrow peak which implies that during the mixing process, the compatibility between butadiene domain of HIPS and SBS is significant. Since SBS is a block copolymer, some of the styrene endblocks can extend into the PPO/HIPS matrix phase, hence, the PPO/HIPS/SBS multiphase blends have enough interfacial adhesion to resist impact damage.

Figure 13 shows the tan δ curves of PPO/HIPS/MBS blends versus temperature with different MBS contents. At low temperature range, two peaks can be found. The first one is around −76°C which is attributed to the rubbery phase of HIPS. The second one occurs around −45°C which comes from the relaxation transition of styrene butadiene components in MBS copolymer. A shift to 5°C higher indicates that a good compatibility between the two phases existed. However, even after adding 20 phr MBS, no further shift can be found which implies that the compatibility is poor. The area of tan δ peak in the low temperature range was reduced after adding MBS modifiers.

Morphology

The SEM micrographs of the fracture surfaces of PPO/HIPS/RUBBER blends after impact test are shown in Figures 14 and 15. From the fracture surfaces of specimens containing 20 phr SBS modifier, one can see those that are compact, regular and white in the broken edge which indicates these blends are ductile in

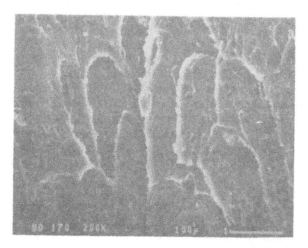

FIGURE 14. The SEM micrographs of the fracture surfaces of PPO/HIPS/SBS (20 phr) blends after impact test.

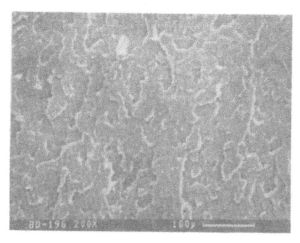

FIGURE 15. The SEM micrographs of the fracture surfaces of PPO/HIPS/MBS (20 phr) blends after impact test.

nature. The fracture surfaces of specimens containing 20 phr MBS modifier are smooth and flat which indicates brittle deformation. The fracture behavior confirms that the adhesive force between MBS modifier and PPO/HIPS is worse than that of SBS modifier.

CONCLUSIONS

1. Adding SBS modifier (≤ 20 phr) to PPO/HIPS matrix will improve the impact strength and elongation of PPO/HIPS blends since a good interfacial adhesion existed between the rubbery phase and PPO/HIPS matrix.
2. Since MBS is stiffer than SBS the decrease in tensile strength and modulus of PPO/HIPS/MBS are less than those of PPO/HIPS/SBS.
3. The addition of either MBS or SBS will not affect the heat distortion temperature, even though 20 phr of rubber modifier were added.
4. From the dynamic mechanical studies, it was found that adding SBS modifier will shift the tan δ peak of the rubber domain of the HIPS resin to 9°C higher, while adding MBS modifier will shift it only 5°C. This result indicates that compatibility between the rubbery phase of HIPS and SBS is better than that of HIPS and MBS modifier.
5. SEM results indicate that the interaction force between the matrix and SBS is much better than that of MBS.

ACKNOWLEDGEMENTS

The authors would like to express their appreciation to the Catalyst Research Center (CRC), China Technical Consultant, Inc. for pemission to publish part of the research results conducted here. Thanks also to Mr. J. S. Ho of CRC, for his assistance in preparing the blends.

REFERENCES

1. Rollmann, K. W., U.S. Patent 4,322,506 (1982).
2. Cizek, E. P., U.S. Patent 3,383,435 (1968).
3. Petrich, R. P. *Polym. Eng. and Sci.,* 13:248 (1973).
4. Manson, J. A. and L. H. Sperling. *Polymer Blends and Composites.* New York and London: Plenum Press. Chapters 2 and 3 (1976).
5. Yoshimura, D. K. and W. D. Richards. *Modern Plastics.* 64 (March, 1987).

BIOGRAPHY

Dr. Chen-Chi Martin Ma

Dr. Chen-Chi Martin Ma has been a Professor of Chemical Engineering at National Tsing Hua University, Hsin-Chu, Taiwan, Republic of China since 1984. He received his Ph.D. degree in Chemical Engineering from North Carolina State University at Raleigh, North Carolina in 1978. He has worked in the past with Monsanto Company (1977–1979) as a senior research engineer and polymer rheologist, with Lord Corporation (1979–1980) as a senior research associate, and with Phillips Petroleum Company (1980–1984) as a research engineer in the Advanced Materials Section at the Phillips Research Center. Professor Ma's research interests include the processing of high performance thermoplastic composites, pultrusion and polymer blends. Professor Ma is an author and coauthor of textbooks and more than 50 technical papers and patents.

7

Polymer Emulsion Blends as Adhesives for Flexible Packaging

JUDITH S. MORPHY,* THOMAS M. SANTOSUSSO*
and DAVID J. ZIMMER*

ABSTRACT: Water-borne adhesive systems typically are based on single polymer types. However, occasionally synergistic effects are encountered where blends of two or more polymers result in improvements in properties over those of each of the polymers individually. In this paper we report the use of three component simplex lattice design experimentation to achieve optimum blends of acrylic, urethane and epoxy emulsions for adhesive applications.

INTRODUCTION

WATER-BORNE LAMINATING ADHESIVES are becoming an increasingly attractive solution for the many environmental, health and safety problems which converters of flexible packaging materials face today. Not the least of these problems is the need to drastically reduce solvent emissions to meet the Environmental Protection Agency's National Ambient Air Quality Standards [1]. In some states, like California, stringent requirements are now in place, with the goal of complying with the Standard by the end of 1987 [2].

Water-based adhesives offer the possibility of reducing or even completely eliminating the use of solvents throughout the laminating process. In addition to reducing direct emissions, this would result in a lessening of fire and worker exposure hazards and waste handling problems [3]. Also, adopting a water-based adhesive often allows a converter to use his existing application and drying equipment, which is not the case for other non-solvent systems such as hot melts, two component systems or radiation cured adhesives [4].

In the best of all possible worlds, polymer chemists would provide adhesive formulators with a single water-borne polymer which would meet all the diverse

* Lord Corporation, Industrial Coatings Division, 2000 West Grandview Blvd., Erie, PA 16514.

performance and application requirements of the converter and would do so without requiring any change in processing equipment or conditions. In real life, however, things are more complicated. In order to balance factors like ease of application, wettability, drying characteristics, bond strength, clarity, environmental resistance and especially cost, it turns out that a blend of two or more polymer emulsions or dispersions is necessary [5]. The usefulness of this approach is indicated by the number of recent literature citations to blends of aqueous dispersions for adhesive and related coatings applications. These aqueous systems include blends of specialty acrylic resins [6]; blends of acrylics with SBR [7], hydrocarbon resins [8], poly(vinylidene chloride) [9], or modified natural rubber [10]; SAN-natural rubber latex blends [11]; poly(chloroprene)-SBR blends [12]; and blends of water-based polyurethanes and PVC copolymers [13,14,15].

The nature of the compromises required when blending water-based adhesives has been pointed out [16]. However, the process of selecting an optimum formulation can be greatly expedited by using appropriate experimental design methodologies. In the case of blends, where the sum of the component fractions must be one, a simplex lattice design can be used [17,18].

In the study described here, three components were used: an acrylic emulsion, an epoxy emulsion, and a urethane dispersion. Preliminary work had indicated that combinations of these systems exhibited a synergy which could be exploited in preparing cost-effective, solvent-free adhesives for flexible films. To optimize these combinations for particular film constructions, a three-component simplex design was chosen which would allow statistically valid prediction of properties over a wide range of compositions with a minimum number of experiments.

MATERIALS AND METHODS

Polymer Latexes

Three polymer emulsions, an acrylic, an epoxy, and a urethane were used in the simplex blending study. Solids were adjusted to 30% w/w with deionized water before blends were made. All latex formulations conform to FDA 175.105.

Films and Laminates

Three basic constructions were prepared using the latex blends as adhesives:

1. Polyester/LLPE
2. Polyester/nylon
3. Al foil/polyester

Laminates were prepared by casting 1.25 lb/ream adhesive films (dry weight) with a wire wound bar. Films were dried in a forced air oven at 72°C for 15 seconds. Laminates were made on a Talboys tabletop laminator operating at 25 ft/min with a nip temperature of 150°F and a pressure of 50 psi. Once laminates were prepared they were stored at 23°C and 55% RH before testing.

Bondability

Laminates were aged and peel adhesion measured by pulling 1 " wide strips at a peel angle of 90° on a Monsanto tensile tester at a speed of 12 "/min. Bond strengths were determined under the following conditions:

(a) Initial (within 1 hr of laminating)
(b) After 24 hours at 23°C
(c) After 24 hours at 50°C
(d) After 7 days at 23°C
(e) After 14 days at 23°C

Wettability

Adhesives were drawn and allowed to dry on a polyester film. The adhesive layer was rated visually for imperfections on a scale of 1–5, 5 indicating a defect free film.

Clarity

Polyester films were laminated and checked visually for clarity. Laminates were graded 1 through 5, higher numbers evidencing greater clarity.

Boilability

Laminates aged 7 days at 23°C were held 30 minutes in boiling water. They were rated pass/fail with any delamination, bubbling, or tunneling rated a failure.

Experimental Design

The experimental design chosen for this study was a three component simplex-centroid utilizing a special cubic polynomial for the regression equation [17]. In

Table 1. Composition of blends for a three component simplex-centroid design.

Experiment Number	Mixture Composition		
	Component 1	Component 2	Component 3
1	1	0	0
2	0	1	0
3	0	0	1
4	1/2	1/2	0
5	1/2	0	1/2
6	0	1/2	1/2
7	1/3	1/3	1/3
8	2/3	1/6	1/6
9	1/6	2/3	1/6
10	1/6	1/6	2/3

this design, seven experimental points in three component mixture space are required to define each response under study. These correspond to the three vertices, three mid-edge points, and the centroid of an equilateral triangle. In addition, three additional points are determined, symmetrically arranged mid-way between the centroid and each of the three vertices. The entire design was replicated. The resulting compositions required are shown in Table 1. The first seven points are used to establish the coefficients of the regression equation, while the remaining three are used to verify the accuracy of the model. This general approach for mixtures has been reviewed [19], and it has been applied specifically to emulsion blends.

The calculation of the coefficients of the regression equation and the lack of fit of the model as indicated by the F ratio, as well as the plotting of the response contours, were carried out using the in-house developed program SIMPLEX.

RESULTS

Responses of the experimental design for the optimized blends for each of the three constructions are shown in Table 2. The optimized blends were selected on the basis of a normalized cumulative score derived from a simplex plot of the sum of the normalized responses. These are shown in Figures 1–3. The values of the responses shown for the optimized blends were calculated on the basis of the regression equation developed for each response. The calculated F ratio for the response in Figures 1 and 2 are smaller than the tabulated value for 3 and 10 degrees of freedom at the .05 probability level. Figure 3 is slightly higher than that same value. It can be concluded that the model within the experimental error of the data is a representation of the response surface of these mixtures.

DISCUSSION

The most important response in this study is naturally the measured bonding ability of the latex blends, and bonding was weighted heavily in the cumulative scores. Laminates were tested in peel within an hour of being made and then after 24 hours, 7 days and 14 days of controlled (23°C, 55% RH) aging. Values after 24 hours at 50°C were also determined. Initial bonds are indicative of the handleability of the construction fresh off the machine. Aging tests were done to determine when ultimate strength was developed.

Much of the bond performance data was specific to individual constructions, but some universal comments can be made. In general, the better initial bonds were given by near equal blends of the three latexes. On aging the polar components of the blends began to exert a stronger influence on adhesion. The adhesive strength of laminate aged overnight at 50°C correlated well with ultimate strength. Ultimate strength plateaued after 7 days aging, as the 7 day and 14 day values were indistinguishable.

The PET/LLPE construction gave the highest absolute bond values with a preponderance of stock tears. The other two laminates evidenced similar though somewhat lower adhesion. Interestingly, the ultimate adhesion values on the PET/nylon laminate never exceeded the initial measurements.

Table 2. Design study responses for optimized blends.

Structure	Polyester to Linear Low Density Polyethylene	Polyester to Nylon	Polyester to Foil
Optimized Blend	36% acrylic	80% acrylic	58% acrylic
	48% epoxy	15% epoxy	32% epoxy
	16% urethane	5% urethane	10% urethane
Initial Bonds	800 gram/inch	600 gram/inch	500 gram/inch
24 Hour Bonds	1000 gram/inch	400 gram/inch	800 gram/inch
24 Hour Elevated Temp. Bonds	1400 gram/inch*	600 gram/inch	570 gram/inch
7 Day	1500 gram/inch*	350 gram/inch	600 gram/inch
14 Day	1500 gram/inch*	200 gram/inch	800 gram/inch
Boilability[a]	3	2	1
Relative Cost[b]	65	40	53
Clarity[c]	3	5	3
Wettability[d]	5	5	5
Humidity	1100 gram/inch	100 gram/inch	200 gram/inch
Normalized Cumulative Score[e]	86	54	56

*Note: If a stock split was observed, a value of 1135 gram/inch was assigned. If a stock tear was observed, a value of 1360 gram/inch was assigned.
[a]Boilability: 0 = fail; 1 = pass; score – sum of 3 trials.
[b]Relative cost: highest cost = 100.
[c]Clarity: rated 1–5; 5 = completely clear, 1 = opaque.
[d]Wettability rated 1–5; 5 = defect free; 1 = no wetting.
[e]Normalized sum of normalized values—maximum 100.

Clarity, wettability, and humidity rated at 0.5 normalized value. All others rated at full value.

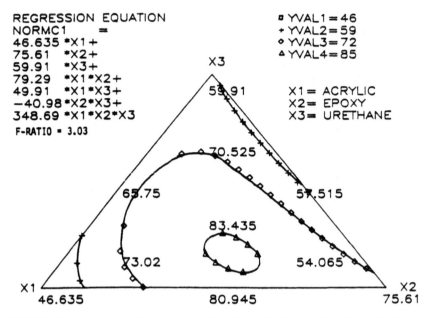

FIGURE 1. Normalized cumulative score of emulsion blends for the polyester to linear low density polyethylene construction.

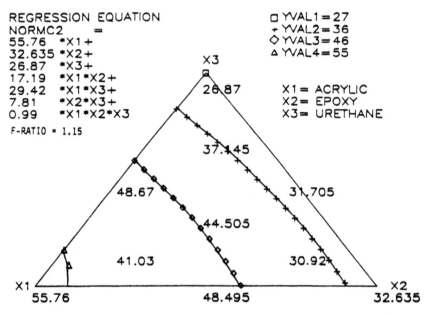

FIGURE 2. Normalized cumulative score of emulsion blends for the polyester to nylon construction.

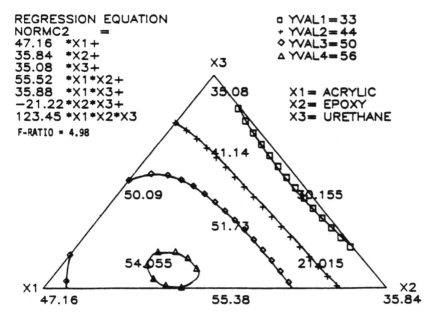

FIGURE 3. Normalized cumulative score of emulsion blends for the polyester to foil construction.

An adhesives viability in the medium to high performance area is dependent on its ability to withstand boiling and humidity conditions. This study indicates that good boilability can be obtained by blending latexes with the major component being the epoxy. An exception to this was the foil/polyester construction which had better results when there is a shift towards the acrylic. In humidity testing the systems were more sensitive to the substrate. A ternary blend yields the highest bond values in a nylon construction. In the other constructions the acrylic and epoxy components have a larger influence on the overall bond performance.

Adhesive film clarity is always a consideration when bonding transparent webs. Polymer latexes range from the familiar milk-like opacity to virtual water transparency depending on the latex particle size. Even opaque latexes generally dry to a transparent film with ultimate clarity dependent on completeness of particle coalescense.

The latexes used in this study singly dry to clear films. In blend, however, there is a loss of clarity, while the urethane and acrylic blend, to give a clear product inclusion of the epoxy with either of the two, results in some loss of transparency. The ternary blend is the worst case.

This study defined wettability as the property of giving a visually defect free film when coated on a PET film. The pure acrylic latex was outstanding in this respect and inclusion of minor amounts of the acrylic improved the flow-out of both the urethane and epoxy. Unexpectedly, the binary blend of the epoxy and urethane rivaled the acrylic.

CONCLUSION

The use of statistically designed experiments based on blending latexes aids in the optimization of adhesive systems for specific laminating constructions. Through this technique, adhesives can be formulated based on the needs of the adhesive application with a minimum amount of trials.

REFERENCES

1. Joseph, R. *Mod. Paint. Coat.*, 77(4):36–38 (April, 1987).
2. Joseph, R. "Getting Into Compliance with California's Air Pollution Laws for Industrial Coatings," Ron Joseph & Associates, Cupertino, CA (1987).
3. Anonymous. "Solvent Limits Proposed Would Cut Exposures," *Chem. Market Rep.*, p. 7 (Jan. 19, 1987).
4. Bashford, R. L. *TAPPI Proceedings—1986 Polymers, Laminations & Coatings Conference.* pp. 25–28.
5. Argent, D. *TAPPI Proceedings—1985 Polymers, Laminations & Coatings Conference.* pp. 255–262.
6. Nakayama, Y., T. Watanabe and I. Toyomoto. *J. Coat. Tech.*, 56(716):73–84 (September, 1984).
7. Marongiu, J., Fr. Demande FR 2,529,899 (January 13, 1984); assigned to Rhône-Poulenc Specialties Chimiques; CA 100, 176179n (1984).
8. Mori, M., M. So and K. Yamaji, Jpn. Kokai Tokkyo Koho JP 61,53,378 (March 17, 1986); assigned to Sunstar Giken K.K.; CA 105, 135088s (1986).
9. Padget, J. C and D. H. McIlrath, Eur. Pat. Appl. EP 119,698 (September 26, 1984); assigned to Imperial Chemical Industries; CA 102, 7918p (1985).
10. Nitto Electric Industrial Co., Jpn. Kokai Tokkyo Koho JP 59,210,984 (November 29, 1984); CA 102, 150587k (1985).
11. Freidin, A. S., M. G. Malyarik, M. M. Belousova, Zn. N. Ospanova and Sh. M. Klauzner, U.S.S.R. SU 1,208,057 (January 30, 1986); assigned to V. A. Kucherenko, Central Scientific-Research Institute of Building Structures; CA 104, 208701c (1986).
12. Sano, S. and R. Tanabe, Jpn. Kokai Tokkyo Koho Jpn. 60,179,480 (September 13, 1985); assigned to Mitsui Toatsu Chemicals; CA 104, 51789n (1986).
13. Narisawa, S., C. Tomizawa, Y. Kanejima and M. Domoto, Jpn. Kokai Tokkyo Koho Jpn. 60,163,981 (August 26, 1985); assigned to Sumitomo Chemical Co.; CA 104, 131048k (1986).
14. Tani, K. and M. Okitsu, Jpn. Kokai Tokkyo Koho Jpn. 60,255,876 (December 17, 1985); assigned to Nippon Urethane Service Co., CA 105, 61755h (1986).
15. Tani, K. and M. Okitsu, Jpn. Kokai Tokkyo Koho Jpn. 60,255,878 (December 17, 1985); assigned to Nippon Urethane Service Co.; CA 105, 25410h (1986).
16. Brooks, T. W., R. M. Kell, L. G. Boss and D. E. Nordhaus. *TAPPI Proceedings—1984 Polymers, Laminations and Coating Conference.* pp. 469–477.
17. Cornell, J. A. *Experiments with Mixtures.* New York:John Wiley and Sons, Chapters 2 and 5 (1981).
18. Hesler, K. K. and J. R. Lofstrom. *J. Coat. Technol.*, 53:33–40 (1981).
19. Snee, R. D. *Chemtech.* 9:702–707 (1984).

8

Determination of the Chain Configuration in Polymer Blends Using Fluorescence Anisotropy Measurement

Y. H. JENG* and C. W. FRANK*

ABSTRACT: An experimental study of polymer structure using electronic excitation transport induced fluorescence depolarization monitored by steady-state anisotropy is presented. We demonstrate that transport of electronic excitations among chromophores randomly attached to a polymer chain can be used as a probe of the configurational properties of an isolated polymer chain in solid-state polymer blends. A transesterification method was developed to attach a small concentration of anthracene groups to the methyl methacrylate repeating units in poly(methyl methacrylate) (PMMA) homopolymer and the Rohm and Haas K125 terpolymer. The K125 which consists of 82% methyl methacrylate, 12% ethyl acrylate and 6% butyl acrylate is used as a commercial processing aid, particularly for poly(vinyl chloride) (PVC). Several polymer blends were examined:

1. PMMA*/PMMA, where the asterisk refers to the anthracene labeled polymer and both the labeled and unlabeled material were the same molecular weight ranging between 12,000 and 96,000
2. K125*/K125, where the K125 has a high molecular weight of 2.2×10^6
3. K125*/PVC, where the PVC has a molecular weight of 133,000

The Flory characteristic ratios for K125*/K125 and K125*/PVC were found to be in the range of 7.5 to 7.9, in good agreement with literature results for syndiotactic PMMA under theta conditions.

INTRODUCTION

ELECTRONIC EXCITATION TRANSPORT (EET) describes the transfer of electronic excitations among chromophores or dye molecules [1]. The transfer process

---

* Department of Chemical Engineering, Stanford University, Stanford, CA 94305.

can proceed through several mechanisms. First, the excitation energy might be released as radiative emission that a nearby chromophore could absorb. Second, this energy could also be transferred nonradiatively through short range exchange interactions induced by wave function overlap of the two chromophores. Finally, long range electrostatic resonance interactions can cause the nonradiative migration of an electronic excitation. In the following discussion we restrict our attention to singlet excitations for which the nonradiative resonant EET mechanism is dominant. The trivial process of radiative transport through emission and reabsorption can be minimized by performing experiments at a low chromophore concentration.

An expression for the rate of singlet excitation transport between two identical chromophores with long range resonant interactions was first derived by Forster [2]. He found that the rate, W, of EET from a chromophore at position $\underline{r}_1$ to a chromophore at $\underline{r}_2$ is approximately given by:

$$W = \frac{1}{\tau}\left(\frac{R_0}{|\underline{r}_1 - \underline{r}_2|}\right)^6 \tag{1}$$

Here the measured lifetime of the electronic excitation is denoted τ, and R_0 is defined as the interchromophore separation distance at which the rate of EET is equal to the lifetime decay rate.

When chromophores are placed in solution or in a solid material at finite concentration, an excitation localized on a particular chromophore can be transferred to any one of a large number of surrounding chromophores. This is a result of the long-ranged transfer rate depending on the inverse sixth power of the chromophore separation distance. A mathematical description of the EET process in such a system will involve the formulation and solution of a many-body transport problem. Pioneering theoretical work by Gochanour, Andersen and Fayer (GAF) [3], and Loring, Andersen and Fayer (LAF) [4] established the understanding of transport of electronic excitations between randomly distributed sites in homogeneous systems. The GAF and LAF theories demonstrate that the strong distance dependence of the microscopic transfer rate will yield a collective or macroscopic rate of EET that is very sensitive to chromophore density, and to fluctuations in chromophore density.

Fredrickson, Andersen and Frank (FAF) [5–10] subsequently built upon the GAF and LAF results in a series of theoretical papers on EET in inhomogeneous polymer systems. They showed that when chromophores are attached to polymer chains in a well-defined manner, the EET dynamics will reflect ensemble density, flexibility, global configuration statistics, and intramolecular dimensions. Although the theory was general, specific calculations were described for an ideal polymer chain having infinite molecular weight. As the number of repeating units of the polymer chain approaches infinity for such a case, all sites on the chain become equivalent and translational invariance along the chain contour may be employed as a mathematical simplification.

Over the same time period that the FAF work was being developed, Ediger and

Fayer (EF) [11] took a more direct approach and developed a finite volume theory to describe EET within a single polymer coil or aggregate. The unique feature of their treatment is that the dynamics of excitation transport depend on the point of initial excitation. An excitation at the center of a sphere encompassing a collection of chromophores will have a different excitation transport rate compared to that for an excitation near the surface of a sphere. This is because of the distinctly different spatial distribution of unexcited chromophores surrounding each of these initial locations. Therefore, the resulting loss of excitation probability from an initially excited chromophore must be computed as an average over the positions of initial excitation. Excitation transport among chromophores randomly tagged at low concentration along a finite length polymer coil can also be analyzed by this finite volume approach.

The experiment that will allow the investigation of macromolecular structure is a fluorescence depolarization study. When a collection of chromophores with randomly oriented absorption dipoles is illuminated with a plane polarized source, chromophores having dipoles aligned with the direction of polarization are preferentially excited. The subsequent fluorescence from these chromophores retains a high degree of polarization. However, if EET moves the excitations to surrounding chromophores with different orientations, the resulting fluorescence will be depolarized.

The extent of fluorescence polarization at a time, t, following a brief excitation pulse is directly related to the time dependent ensemble averaged probability that an excitation resides on the chromophore where it was created at $t = 0$ [3,4]. This probability, which will be designated $G^s(t)$, depends strongly on the density and distribution of chromophores, and can be obtained from a theoretical analysis of the many-body EET problem. All of the time dependent and photostationary fluorescence observables can be obtained from the Laplace transform of $G^s(t)$, which is denoted $\hat{G}^s(\epsilon)$. Several experiments [12,13] using time-resolved fluorescence depolarization measurements have confirmed the theories and have given a detailed description of excited-state transport in the system.

Gochanour and Fayer [12] investigated EET for solutions of rhodamine 6G (R6G) in glycerol. The time-resolved fluorescence depolarization measurements yielded an R_0 value of 50 Å for R6G which is similar to the 47 Å obtained by the spectral overlap method developed by Forster [2] and used by Kawski [14]. These results confirm the accuracy of the theoretical method [3] for the comprehensive description of EET among chromophores in solution. Ediger et al. [13] examined copolymers of methyl methacrylate and 2-vinylnaphthalene (2VN) (molecular weight = 20,000 and 2VN mole fraction = 0.1) in poly(methyl methacrylate) hosts by electronic excitation transport induced fluorescence depolarization. Quantitative determination of isolated polymer coil size through the root-mean-square radius of gyration was compared with recent theoretical treatments [5,11].

The FAF theory uses an approximation that is best applicable only to infinite length chains. When applied to a coil of finite size it overestimates the local chromophore density, causing the theory to yield an average chain configuration that is too large. Conversely, the EF theory underestimates the local chromophore density, leading to a determination of the average coil size that is too small. Recently, a theory based on a first order cumulant expansion developed by Huber [15] has been applied to describe excitation transport on isolated, finite size, flexible polymer coils by Peterson and Fayer [16]. This method has the advantage of being a relatively straightforward time-domain calculation, instead of being in Laplace space, making the inclusion of different chromophore distribution functions mathematically tractable.

All of the preceding theoretical and experimental work has emphasized use of the transient anisotropy. For the present study, we perform the appropriate integration to obtain predictions about polymer chain structure from steady-state anisotropy measurements. From knowledge of the photophysical parameters, we may then determine the value of Flory's characteristic ratio, C_∞, which is a measure of the statistical chain flexibility. Flory [17] proposed that the structure of bulk amorphous polymers is random in nature, and that the configurational statistics of a particular chain molecule is the same as in a θ solvent. In the present study, the statistics of the isolated chain were studied in three different amorphous polymer blends. The first polymer system of interest is the Rohm and Haas K125 terpolymer which has the composition of 82% methyl methacrylate, 12% ethyl acrylate and 6% butyl acrylate. In addition, we have examined the K125/PVC blend and several poly(methyl methacrylate) (PMMA) samples of varying molecular weight. The characteristic ratios of the labeled polymer chain in these amorphous polymer blends were compared to those obtained in a θ solvent.

In order to examine these inherently nonfluorescent polymers, a transesterification method was developed to attach a small concentration of anthracene groups to the methyl methacrylate repeating units. There are relatively few papers published describing the labeling of anthracene luminescent markers onto PMMA chains. Most are based upon the initial synthesis of a monomer containing a fluorescent group followed by free radical copolymerization with methyl methacrylate [18–20]. Recently, however, Teyssie et al. [21] have modified high molecular weight PMMA by adding a 9-methyl anthryl lithium organometallic compound to a dry THF solution of PMMA. Through this approach they were successful in labeling anthracene to the carbonyl of the ester groups. The approach taken in the present study is similar to that of Teyssie in that we begin with a well characterized nonfluorescent polymer and then chemically modify it so that it might be used as a fluorescent probe polymer.

EXPERIMENTAL

Synthesis

Nearly monodisperse PMMA samples having number average molecular weights of 26,000, 29,700, 65,000 and 125,000 were obtained from Pressure Chemical Company. In addition, a broad molecular weight distribution PMMA sample with $M_n \sim 12,000$ was also obtained from Pressure Chemical Company.

The K125 terpolymer, which is a Rohm and Haas product consisting of 82% methyl methacrylate, 12% ethyl acrylate and 6% butyl acrylate, was a gift from Dr. Wendel Schuely of Aberdeen Proving Ground. The 9-anthracene methanol, transesterification catalyst $Mg(OAc)_2 \cdot 4H_2O$, and solvents N,N-dimethylformamide (DMF) and N,N-diethylformamide (DEF) were obtained from Aldrich Chemical Company. All materials were used as received.

The reaction involved the direct transesterification between the methacrylate ester groups of the polymer chain of PMMA and 9-anthracene methanol. In a typical reaction the PMMA (0.93–2 g), 9-anthracene methanol (0.33–1.0 g), $Mg(OAc)_2 \cdot 4H_2O$ (0.04–0.08 g) and dimethylformamide or diethylformamide (10–20 ml) were placed in a three-necked flask. Here the numbers in parentheses represent the ranges of quantities examined. The mixture was heated to 426–450 K and kept at this temperature for 10–120 min. The reaction was run under reflux conditions with the temperature established by the choice of solvent; the boiling points of DMF and DEF are 426 K and 450 K, respectively. The labeled polymer was purified of excess 9-anthracene methanol by repeated acetone/ethanol or benzene/water solution precipitation until the UV absorbance of anthracene remained constant. Gel permeation chromatography measurements were made with a Model 244 Waters Associates Liquid Chromatography system with dual detectors (refractometer and UV absorbance at 313 nm) and six microstyragel columns of 10^6 Å, 10^5 Å, 10^4 Å, 10^3 Å, 500 Å and 100 Å. Based on the presence of a strong UV absorbance peak after the reaction, it was clear that anthracene chromophores had been introduced into the polymer chain.

Spectroscopy

Blends were prepared by first dissolving 1 wt% of the labeled polymer along with 99 wt% of the matrix material in tetrahydrofuran solvent. The labeled material was either K125* or PMMA*, while the matrix was K125, PMMA or PVC. Solid films for spectroscopic study were prepared by casting this solution onto glass microscope slides, after which the slides were kept in a vacuum oven for several days at 313 K. The fluorescence spectra of these films were taken with a Spex Fluorolog 212 having a 450 W Xenon lamp and Hammamatsu R928 photomultiplier. Two UV Glan-Thompson polarizers were installed, one for initial polarization of the exciting light before it reached the sample and one for polarization of the emission light that was passed on to the emission spectrometer.

The important experimental parameter is the emission anisotropy, r, which can be calculated from the relation:

$$r = (I_{VV} - GI_{VH})/(I_{VV} + 2GI_{VH}) \qquad (2)$$

where I is the intensity of emitted light with suffixes V and H denoting vertically and horizontally polarized light. The first suffix corresponds to the incident light and the second to the emitted light. The $G(= I_{HV}/I_{HH})$ factor in Equation (2) is necessary to correct for the depolarization characteristics of the apparatus.

Table 1. Summary of anthracene labeling percentage as functions of reaction time and temperature.

Sample No.	Polymer Type	Molecular Weight	Reaction Temp., K	Reaction Time, min	Anthracene Labeling, %
1	PMMA	$\overline{M}_n$ = 12,000	426	106	0.25
2		by light	426	137	0.25
3		scattering	449	10	0.057
4			449	20	0.2
5			449	30	0.4
6			449	40	0.7
7			449	50	1.25
8			449	60	1.86
9			457	137	1.1
10	K125	$\overline{M}_w$ = 2,200,000	449	54	0.14
11			449	66	0.41
12			449	120	1.65
13	PMMA	26,000, $\overline{M}_w/\overline{M}_n$ = 1.08	426	60	0.026
14			449	30	0.64
15			449	80	0.66
16	PMMA	29,700, $\overline{M}_w/\overline{M}_n$ = 1.07	449	15	0.21
17		65,000, $\overline{M}_w/\overline{M}_n$ = 1.09	449	40	0.36
18		96,000, $\overline{M}_w/\overline{M}_n$ = 1.10	449	136	0.53
19			449	175	0.86
20	PMMA	125,000, $\overline{M}_w/\overline{M}_n$ = 1.08	449	50	0.21
21			457	50	0.1

RESULTS

Synthesis of Anthracene Labeled Polymers

The reaction temperature, reaction time and percentage of pendant anthracene attached to the PMMA or K125 are tabulated in Table 1. As the reaction time is increased, the anthracene tagging percentage is also increased, as shown in Figure 1. It also appears that there is an induction period for the K125 labeling reaction. This probably arises from the difference in molecular weight with the K125 being over 200 times greater than that for the PMMA. Thus, one might expect diffusional limitations, particularly since the polymer solution concentration was relatively high, approximately 10% by weight. The extent of labeling also increases with an increase in the reaction temperature, as shown in Figure 2. It was necessary to limit the temperature range, however, because if the temperature were to exceed 460 K, the pendant anthracene would start to decompose thermally.

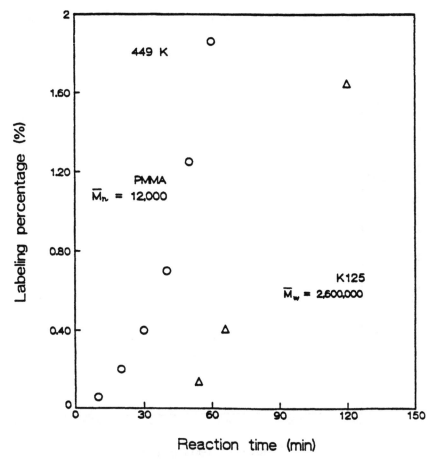

FIGURE 1. Effect of reaction time on anthracene tagging percentage for reaction at 449 K. ○ = PMMA homopolymer with $\overline{M}_n$ = 12,000; △ = K125 terpolymer with $\overline{M}_w$ = 2,200,000.

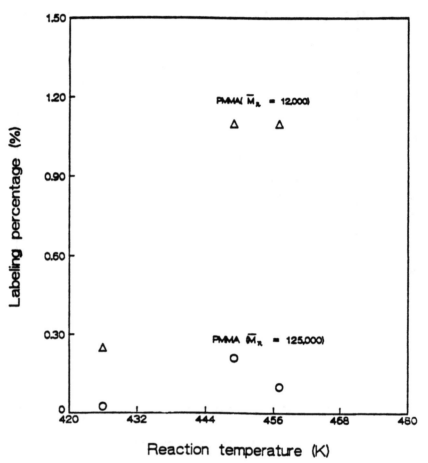

FIGURE 2. Effect of reaction temperature on anthracene tagging percentage. $\bigcirc$ = PMMA homopolymer with $\overline{M}_n$ = 12,000; $\triangle$ = K125 terpolymer with $\overline{M}_n$ = 125,000.

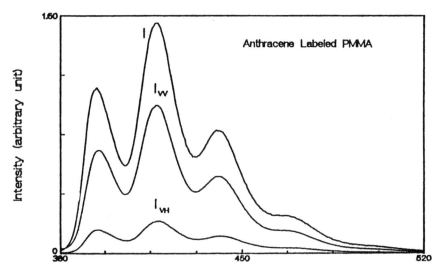

FIGURE 3. Polarized and unpolarized fluorescence spectra of anthracene labeled PMMA.

Excitation Transfer in K125 Terpolymer and PMMA Homopolymer

A plot of the polarized fluorescence spectrum of anthracene labeled K125 is shown in Figure 3. In this figure, curve I_{VV} shows the emission fluorescence intensity for the case where the polarization vectors of the two polarizers are parallel; curve I_{VH} is the measured fluorescence intensity when the polarization vectors of the two polarizers are perpendicular and curve I is the anthracene emission intensity. The measured anisotropies are tabulated in Table 2.

Table 2. Summary of characteristic ratio C_∞
for PMMA*/PMMA, K125*/K125 and K125*/PVC blends.

Sample	$\bar{M}_n$	$\bar{M}_w$	$\bar{M}_w/\bar{M}_n$	mol. % tag	$\bar{r}/0.4$	C_∞
PMMA*/PMMA	29,700	31,779	1.07	0.21	0.95	10.1
	26,000	28,080	1.08	0.66	0.86	9.5
	96,000	105,600	1.10	0.82	0.84	9.8
	12,000			1.86	0.71	10.7
K125*/K125	500,000	2,200,000		0.41	0.89	7.52
				1.65	0.65	7.89
K125*/PVC				1.65	0.62	7.62

DISCUSSION

Electronic Excitation Transport Theory

In this paper we are generally concerned with an isolated chain having a small fraction of the repeating units containing chromophores that is dispersed in a solid matrix of the identical, but unlabeled polymer or copolymer. Examples of this are the PMMA*/PMMA and K125*/K125 blends. The locations of the chromophores on the labeled polymer are characterized by the chain position and configuration, $\{R\} = (r_1, \ldots, r_n)$. The probability that an excitation resides on the jth chromophore at time t, $P_j(\{R\},t)$, depends on the chromophore locations $\{R\}$ and satisfies the master equation [5]:

$$\frac{d}{dt} - P_j(\{R\},t) = - P_j(\{R\},t)/\tau + \Sigma_k W_{jk}[P_k(\{R\},t) - P_j(\{R\},t)] \tag{3}$$

where τ is the measured lifetime and W_{jk} is the transfer rate between sites j and k. By introducing the Laplace transform, a Green function for this process can be defined as the ensemble-averaged quantity:

$$\hat{G}(\epsilon) = (\epsilon I - Q)^{-1} \tag{4}$$

in which the transfer matrix is defined by:

$$Q_{jk} = W_{jk} - \delta_{jk}\Sigma_l W_{kl} \tag{5}$$

The quantity most useful for obtaining information concerning transport in a fluorescence depolarization experiment is related to the diagonal elements of the Green function $\hat{G}^s(\epsilon)$. Here $\hat{G}^s(\epsilon)$ is the Laplace transform of the probability that an excitation resides on the site where it was created at time t, $G^s(t)$. Thus, we can analyze the time dependence of polarized fluorescence by dividing the excited molecules in the sample at time t into two distinct ensembles. The first, with a weighting factor $G^s(t)$, consists of molecules initially excited for which the resulting fluorescence is polarized. The second ensemble, with a weighting factor $1 - G^s(t)$, consists of molecules to which excited state energy has been transferred. The fluorescence from these sites is unpolarized.

The relationships between $G^s(t)$ and the observables in the photostationary fluorescence depolarization experiment are [12]:

$$I_\parallel(t) = \exp{(-t/\tau)}[1 + 0.8G^s(t)] \tag{6}$$

$$I_\perp(t) = \exp{(-t/\tau)}[1 + 0.4G^s(t)] \tag{7}$$

where $I_\parallel$ is the fluorescence intensity polarized parallel to the excitation polarization and $I_\perp$ is the fluorescence intensity polarized perpendicular to the excitation polarization. These expressions are valid when the depolarization of fluorescence results only from resonant excitation transfer and not from molecular rotation. The steady-state anisotropy in a photostationary fluorescence depolarization experiment is defined by:

$$\bar{r}(t) = \frac{\int_0^\infty r(t)\, I(t)\, dt}{\int_0^\infty I(t)\, dt} \tag{8}$$

in which the total intensity, $I(t)$, is proportional to $I_\parallel(t) + 2I_\perp(t)$ and $r(t)$ is the transient anisotropy which contains the polarization information. This steady-state anisotropy is directly related to the Laplace transform of $G^s(t)$ [5]:

$$\frac{\bar{r}}{r_0} = \epsilon \hat{G}^s(\epsilon) \tag{9}$$

in which $\bar{r}_0 = 0.4$.

In order to obtain information on polymer structure by using photostationary state anisotropy measurements, an expression for $G^s(t)$ is needed. Under the assumption of isotropic Forster transfer on an infinite polymer chain with ideal Gaussian statistics and containing a small concentration of randomly placed chromophores, Fredrickson, Andersen, and Frank developed a three-dimensional model of intersite transport [5]. According to the FAF three-particle Pade approximant, the Laplace transform of $G^s(t)$ can be expressed as:

$$\hat{G}^s(\epsilon) = \epsilon^{-1} \left\{ 1 + \frac{2^{3/4}\pi}{3^{1/2}} \left[\frac{r_2^3 q^3}{\epsilon \tau} \left(\frac{R_0}{a} \right)^6 \right]^{1/3} + 4.123 \left[\frac{r_2^3 q^3}{\epsilon \tau} \left(\frac{R_0}{a} \right)^6 \right]^{2/3} \right\}^{-1} \tag{10}$$

where R_0 is the Forster radius, a is the statistical segment length of an ideal chain [22], q is the average number of chromophores per segment and $r_2 = 0.8468$. Hence, the expression for the steady-state anisotropy is given by:

$$\frac{\bar{r}}{r_0} = \{1 + 1.4548\, r_2 \bar{c}_D + 0.418\, r_2^2 \bar{c}_D\}^{-1} \tag{11}$$

where the dimensionless concentration is given by:

$$\bar{c}_D = \pi q \left(\frac{R_0^2}{a} \right)^2 \tag{12}$$

$\bar{c}_D$ contains information about chain flexibility through the characteristic ratio

C_∞ [21]. The dimensionless concentration can be expressed as:

$$\bar{c}_D = (\pi\sigma R_0^2)/(C_\infty l^2) \tag{13}$$

where σ is determined synthetically by the extent of labeling and l is the length of a monomer unit. Thus, a single parameter fit of steady-state anisotropy data can provide a value of the characteristic ratio C_∞. The steady-state anisotropy has been plotted in Figure 4 as a function of $\bar{c}_D$. This shows that as the labeling content increases or the chain stiffness decreases, it will be more likely that fluorescence depolarization will occur due to excitation transfer.

Photostationary fluorescence anisotropy measurements can also be used in conjunction with the transient theories [7,9,16] previously developed for excitation transfer between different polymer chains or between fluorescent groups located on the ends of the same chain. We first consider EET among chromophores randomly attached to the polymer chain, with the labeled polymer at high concentration. As the labeled polymer concentration increases, intramolecular and intermolecular transfer become competitive. Using the three-particle Pade approximation for the special case of Forster transfer on ideal chains [9], we find the steady-state anisotropy to be given by:

$$\bar{r}/\bar{r}_0 = \{1 + 1.455 r_2\bar{c}_D + 1.111 r_2 c_D + 0.4176(r_2\bar{c}_D)^2$$
$$\tag{14}$$
$$+ 0.4804 r_2 r_3 c_D\bar{c}_D + 0.4281 r_3 c_D)^2\}^{-1}$$

where the dimensionless bulk chromophore concentration is defined as $c_D = 4\pi/3(R_0)^3\varrho$, ϱ is the bulk number density of molecules and $r_3 = 0.8452$ [9]. The steady-state anisotropy has been plotted in Figure 5 as functions of c_D and $\bar{c}_D$. For low to moderate bulk concentrations (between $c_D = 10^{-2}$ and $c_D = 10^{-1}$) interchain transport is relatively unimportant, but for bulk concentrations greater than $c_D = 10^{-1}$, interchain transport becomes more important. The effect of interchain EET becomes totally dominant at the highest bulk concentration. As interchain excitation transfer becomes more competitive, a smaller anisotropy results due to the accessibility of more excitation transfer pathways. The major difference among curves A, B and C in Figure 5 is due to the interchain transport.

Second, if both ends of the polymer are labeled with chromophores, the end-to-end distance can also be measured through the steady-state depolarization experiment. This measurement may be of particular interest for determining coil size in viscous solution or in the solid-state. In the case of an ideal Gaussian chain for which both chain ends are labeled with the same chromophore, the approximants for $\hat{G}(\epsilon)$ have been obtained in [9], and the steady-state anisotropy becomes:

$$\bar{r}/\bar{r}_0 = 1 - \epsilon\Delta_{11} \tag{15}$$

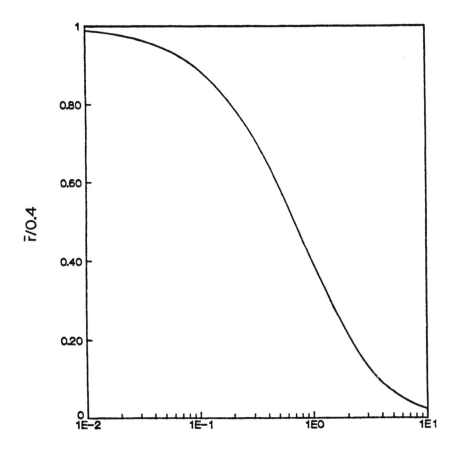

$$\overline{C}_D = \pi \left(\frac{R_0}{\ell}\right)^2 \frac{\sigma}{C_\infty}$$

FIGURE 4. Normalized steady-state anisotropy as a function of $\overline{c}_D$, defined as $\pi\sigma/C_\infty(R_0/l)^2$, for intramolecular EET among chromophores randomly labeled along the polymer chain. The calculation is based on Equation (11).

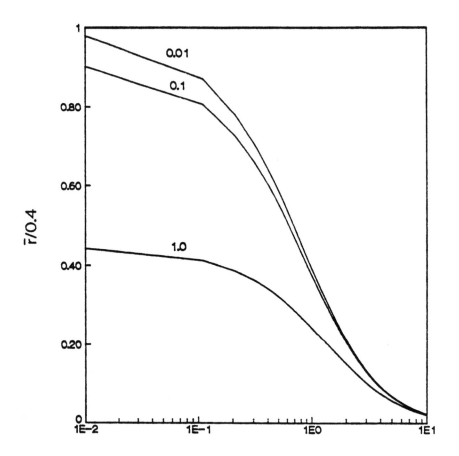

$$\overline{C}_D = \pi q^2 \left(\frac{R_0}{a}\right)^2$$

FIGURE 5. Normalized steady-state anisotropy as a function of $\overline{c}_D$ for intermolecular EET among chromophores randomly labeled along the polymer chain. The numbers by the curves refer to the values of c_D, which is defined as $4\pi/3(R_0)^3\varrho$. The calculation is based on Equation (14).

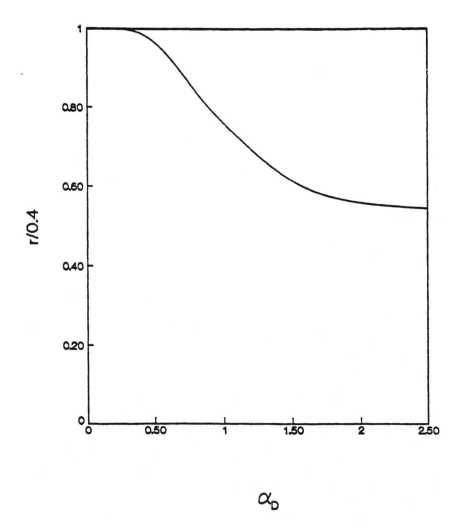

FIGURE 6. Normalized steady-state anisotropy as a function of α_D, defined as $R_0/<R^2>^{1/2}$, for intramolecular EET among chromophores labeled at the chain ends. The calculation is based on Equation (15).

95

where

$$\Delta_{11} = \frac{1}{2} \int_0^\infty 4\pi r^2 dr \; \frac{3}{2\pi <R^2>} \; e^{-3r^2/2<R^2>} \; \frac{1/\tau \; (r/R)^6}{1 + 2/\tau \; (r/R)^6 1/\epsilon} \tag{16}$$

The steady-state anisotropy is plotted in Figure 6 as a function of α_D, defined as $R_0/<R^2>^{1/2}$. The utility of this plot is clear. If we can label different chromophores such as naphthalene, anthracene, or pyrene at both ends of a polymer chain and perform the steady-state fluorescence anisotropy measurement, we can obtain the end-to-end distance from the master curve in Figure 6. The result indicates that as the polymer chain becomes shorter or the Forster radius of the fluorescence probe increases (recall that the excitation transfer rate depends on the sixth power of the probe's Forster radius), then the excitation transfer rate will be enhanced and the anisotropy will decrease.

Evaluation of the Characteristic Ratio

A common objective in polymer physics is the determination of the size and spatial arrangements of a long polymer chain. If the fluorescent techniques proposed by FAF and EF and the present work are to be used effectively, we must first verify that the labeling procedure does not perturb the chain statistics. In this paper we have started with a well characterized, non-fluorescent polymer and then chemically modified it so that fluorescence may be used to study its structure. This has the advantage over the copolymerization approach of retaining the structural features of the initial polymers, particularly the monodisperse character and stereoregularity. The potential disadvantage is the one that is of concern in all labeling experiments—does the covalently bound fluorescent chromophore perturb the structure or dynamic phenomenon under investigation?

We note that Peterson et al. [23] have studied excitation transport among naphthyl chromophores in low concentration on isolated coils of poly(2-vinylnaphthalene-co-methyl methacrylate) in a poly(methyl methacrylate) host. Their results indicate that the presence of the naphthalene-containing monomers does not significantly perturb the average coil dimensions, at least up to a naphthalene label concentration of 9 mol percentage. Since the PMMA homopolymer and K125 copolymer examined in the present study were tagged at a maximum level of 1.86%, with the majority of samples at considerably lower levels, we believe that the fluorescence results are fully representative of the isolated chain.

By using the results of the well-known, random-walk calculation in three dimensions, the root-mean-square, end-to-end distance can be found. Due to steric effects and fixed bond angles, the dimensions of the coiled chain are expanded compared to the freely-jointed model. The unperturbed dimension of the polymer chain can be calculated by taking into account the effects of the geometry and chemical constitution of the polymer molecule but not other factors such as interactions with solvent molecules. A representation of the stiffness of a polymer chain is provided by the characteristic ratio, C_∞, which is the ratio of the unperturbed dimension of the polymer chain to that of the freely-jointed chain. The characteristic ratio can be obtained from measurements under θ conditions [24].

In order to evaluate the significance of the proposed fluorescence technique, it is necessary to have some independent experimental data on C_∞ for comparison. We note that C_∞ has been found to be about 9 in light scattering experiments for atactic PMMA solutions under θ conditions [24]. Neutron scattering measurements reveal that the effect of long-range interactions is small for an amorphous polymer in the solid-state and therefore this condition is similar to that of a θ solvent [17]. Since the solid-state blends of anthracene labeled PMMA in pure PMMA should provide for θ conditions, they would be a reasonable reference point.

We first consider the PMMA*/PMMA blends and note that all of the C_∞ values in Table 2 are larger than the expected value of 9. The probable reason for this is that the PMMA samples under study do not have sufficiently high molecular weight such that the infinite chain assumption and Debye pair correlation function used in the numerical calculation might not be suitable. If the polymer chain does not obey Gaussian chain statistics, the infinite chain model overestimates the local chromophore concentration and, hence, overestimates $<R^2>^{1/2}$. Therefore, the chain will behave as if it were more expanded and the C_∞ values, which are a measurement of the chain size, will be overestimated. These results are consistent with those of Ediger and Fayer [11] who found $<R_g^2>^{1/2}$ to be 50 Å by fitting FAF theory to the transient depolarization experimental data whereas a wide variety of experimental techniques [24] yield a value of 37 $\pm$ 4 Å for a 20,000 molecular weight PMMA homopolymer.

For nonideal chain statistics, the appropriate pair correlation function can be obtained by a two-particle cumulative, approximate method [10] from a transient depolarization experiment. In general, for a finite length chain, the correlation function may be quite complicated and numerical integration is needed in order to obtain $G'(t)$ [16]. Hence, the integration process in Equation (8) will be very difficult to perform to obtain the steady-state anisotropy information. In this case, the transient depolarization measurement is required.

Next we consider the K125 copolymer whose primary commercial use is as a processing aid in the extrusion of polyvinyl chloride. It has also been evaluated as a viscoelastic additive to prevent aerodynamic breakup of liquids. An extensive NMR study [25] has shown the tacticity of methyl-methacrylate in the K125 sample to be predominantly syndiotactic. The C_∞ values lie between 7.3 and 8.4 for predominantly syndiotactic PMMA chains [26–30]. Since the K125 sample used contains 82% methyl methacrylate monomer units, we expect the chain flexibility to behave in a manner similar to that of pure syndiotactic PMMA.

It is necessary, however, to consider whether the presence of even the small amount of ethyl acrylate and butyl acrylate would significantly affect the K125 chain statistics. To evaluate this point, we have examined the data for the solution viscosity of PMMA and the K125 sample. The relationship between intrinsic viscosity, $[\eta]$, of pure PMMA in ethyl acetate solvent and its viscosity molecular weight, M_v, is $[\eta] = 2.11 \times 10^{-4} M_v$. This equation is used as an approximation to PMMA in diethylmalonate solvent because diethylmalonate and ethyl acetate have similar solvent parameters (solubility parameter, hydrogen bond parameter and dipole moment). For PMMA with $M_v = 2.4 \times 10^6$, the intrinsic viscosity is 2.5 compared to $[\eta] = 2.3$ for the same molecular weight of K125 sample. Since the intrinsic viscosity is proportional to the polymer coil unperturbed dimensions $<R^2>^{1.5}$, this value supports our assumption that the chain flexibility of K125 is similar to that of syndiotactic PMMA. The experimental data show that they are consistent with these published values.

Finally, we consider the K125/PVC blend consisting of 1 wt% K125 and 99 wt% PVC. Schurer et al. [31] have employed differential scanning calorimetry to show that compatible blends with PVC having $M_v = 55,000$ were obtained for up to 60 wt% of syndiotactic PMMA (with $M_v = 370,000$). Since the majority of the K125 terpolymer is composed of syndiotactic PMMA, we expect that the K125 terpolymer will form a compatible blend with PVC in our study, at least at low K125 concentration. The characteristic ratio implies that the configuration of the K125 terpolymer chain in this blend is comparable to that in the K125 terpolymer matrix. This implication suggests that the K125 exists in a theta condition.

SUMMARY

Although neutron scattering is quite powerful, the technique does have some limitations including the necessity of deuteration which may affect the thermodynamic or kinetic properties of the polymeric system and the restricted access to facilities. These conditions have motivated our search for fluorescence techniques to monitor the structure of polymer blends in the solid-state. In this study, we have applied electronic excitation transport experiments to the investigation of polymer blends. We have shown that when the chromophore-containing polymer concentration is low, stationary fluorescence anisotropy measurements provide a sensitive tool to study the configurational statistics of an isolated guest coil.

ACKNOWLEDGEMENT

We appreciate the gift of K125 by Dr. Wendel Schuely of Aberdeen Proving Ground. This work was supported by the Polymers Program of the National Science Foundation through Grant DMR-84-07847.

REFERENCES

1. (a) Steinberg, I. Z. *Annu. Rev. Biochem.*, 40:83 (1971); (b) Grinvald, A., E. Haas and I. Z. Steinberg. *Proc. Natl. Acad. Sci.*, 69:972 (1972); (c) Stryer, L. *Annu. Rev. Biochem.*, 47:819 (1978).
2. Forster, Th. *Ann. Phys. (Leipzig)*, 2:55 (1948).
3. Gochanour, C. R., H. C. Andersen and M. D. Fayer. *J. Chem. Phys.*, 76:2015 (1979).
4. Loring, R. F., H. C. Andersen and M. D. Fayer. *J. Chem. Phys.*, 76:2015 (1982).
5. Fredrickson, G. H., H. C. Andersen and C. W. Frank. *Macromolecules*, 16:145 (1983).
6. Fredrickson, G. H. and C. W. Frank. *Macromolecules*, 16:1198 (1983).
7. Fredrickson, G. H., H. C. Andersen and C. W. Frank. *J. Chem. Phys.*, 79:3572 (1983).

8. Fredrickson, G. H., H. C. Andersen and C. W. Frank. *Macromolecules,* 17:54 (1984).

9. Fredrickson, G. H. and C. W. Frank. *Macromolecules,* 17:54 (1984).

10. Fredrickson, G. H., H. C. Andersen and C. W. Frank. *J. Polym. Sci. Poly. Phys. Ed.,* 23:591 (1985).

11. Ediger, M. D. and M. D. Fayer. *Macromolecules,* 16:1839 (1983).

12. Gochanour, C. R. and M. D. Fayer. *J. Phys. Chem.,* 85:1979 (1981)

13. Ediger, M. D., R. P. Domingue, K. A. Petersen and M. D. Fayer. *Macromolecules,* 18:1182 (1985).

14. Kawski, A. Z. *Naturforsch.,* 18:961 (1963).

15. (a) Huber, D. L., *Phys. Rev. B.,* 20(6):2307 (1979); (b) Huber, D. L. *Phys. Rev. B.,* 20(12):5333 (1979).

16. Petersen, K. A. and M. D. Fayer. *J. Chem. Phys.,* 85:470 (1986).

17. Flory, P. J. *Chem. Phys.,* 17:303 (1949).

18. Sheiekhov, N. S., M. G. Krakovyak and S. I. Klenin. *Polymer Science U.S.S.R., Part A,* 1818 (1977).

19. U.S. Patent 3361726, *Chemical Abstract,* 68:40449 (1968).

20. U.S. Patent 3316087, *Chemical Abstract,* 68:70259 (1968).

21. Albert, B., R. Jermoe and P. Teyssie. *J. of Polymer Science Part A: Polymer Chemistry,* 24:537 (1986).

22. Flory, P. J. *Statistical Mechanics of Chain Molecules.* New York:Wiley-Interscience (1969).

23. Peterson, K. A., M. B. Zimmt, S. Linse. R. P. Domingue and M. D. Fayer. *Macromolecules,* 20:168 (1987).

24. Brandrup, J. and E. H. Immergut, eds. *Polymer Handbook.* New York:Wiley-Interscience (1975).

25. Bunting, W. W. and W. J. Shuely. Technical Report ARCSL-TR-8201 (May, 1983).

26. Sajyrada, I., A. Nakajima, O. Yoshizaki and K. Naakamae. *Kolloid,* 41:186 (1962).

27. Schulz, G. V. and R. Z. Kirste. *Phys. Chem. (Frankfurt am Main),* 30:171 (1961).

28. Schulz, G. V., W. Wunderlich and R. Kirste. *Makromol. Chem.,* 22:75 (1964).

29. Sundararajan, P. R. and P. J. Flory. *J. Am. Chem. Soc.,* 96:5025 (1974).

30. Bates, F. S. and G. D. Wignall. *Macromolecules,* 19:932 (1986).

31. Schurer, J. W., A. de Boer and G. Challa. *Polymer,* 16:201 (1975).

BIOGRAPHY

Yoeng Huie Jeng

Dr. Jeng received his B.S. degree in chemical engineering from National Taiwan University in 1980 and his Ph.D. degree in chemical engineering from Stanford University in 1988. He is currently in the metal/polymer interface science group at the IBM T. J. Watson Research Center, Yorktown Heights.

9

Miscible Blends of
Polybutadiene and Polyisoprene

C. MICHAEL ROLAND* and CRAIG A. TRASK**

INTRODUCTION

THE FREE ENERGY change (excess free energy) accompanying formation of a two component mixture can be expressed as:

$$\Delta G_M/kT = V[\phi_i \ln \phi_i /V_iN_i + \phi_j \ln \phi_j/V_jN_j + X\phi_i\phi_j] \quad (1)$$

where V is the total volume, and V_i, N_i and ϕ_i represent respectively the molar volume, degree of polymerization and the volume fraction of the ith component [1–3]. The first two terms on the right hand side of Equation (1) correspond to the ideal entropy, which assumes strictly random mixing. The third term represents, in the simple Flory-Huggins theory of polymer solutions and blends, the mixing enthalpy. The proportionality of the enthalpy to the product of the component concentrations reflects the assumption of random mixing. The interaction parameter, X, is a measure of the energy change associated with replacing the like contacts with unlike contacts:

$$X = (E_{ij} - E_{ii}/2 - E_{jj}/2)/kT \quad (2)$$

In the Flory-Huggins model, this exchange energy is assumed independent of concentration, and the number of like and unlike contacts is only a function of the concentration.

Miscibility implies that a lower free energy is associated with molecular dispersion of the components than with a phase separated morphology. A negative free energy change, however, is not sufficient to ensure stability to composition

*Naval Research Laboratory, Chemistry Division, Code 6120, Washington, DC 20375-5000.
**Geo-Centers, Inc., Fort Washington, MD 20744.

fluctuations over the entire composition range. The spinodal curve, which defines this local stability limit, is obtained by setting the second derivative of the free energy with respect to composition equal to zero. From Equation (1):

$$X_{sp} = V_r/2 \left[1/(V_i N_i \phi_i) + 1/(V_j N_j \phi_j) \right] \tag{3}$$

where the ϕ now refer to the respective concentrations at a given spinodal temperature and V_r is an arbitrary reference volume conveniently taken to be the root mean square of the respective molar volumes of the component chain units. The spinodal point, or minimum on the spinodal curve, defines a value of the interaction parameter below which the system is miscible for all concentrations of the components. The composition at the critical point is deduced by equating the third derivative of the free energy with respect to concentration to zero. For a Flory-Huggins mixture this critical composition is given by:

$$\phi_i^* = (V_j N_j)^{1/2}/[(V_i N_i^{1/2}) + (V_j N_j^{1/2})] \tag{4}$$

and, by inserting ϕ^* in Equation (3), the corresponding critical value of the interaction parameter is obtained:

$$X_{cr} = V_r/2 \left[(V_i N_i)^{-1/2} + (V_j N_j)^{-1/2} \right]^2 \tag{5}$$

It is well-known that the concentration independence of the interaction parameter required by the Flory-Huggins model is rarely observed in polymer mixtures when chemical reaction transpires between the components [4,5]. It has recently been reported that the theory also fails to accurately describe the thermodynamics of mixtures of polymer isotopes [6]. Such experimental results are consistent with theoretical predictions that the number of contacts between chain units depends upon the interaction energy [7,8], contrary to the random mixing assumption of the Flory-Huggins model. Local correlation effects, due to chain connectivity and the attractive forces between like contacts, increase the concentration of like contacts at the expense of unlike contacts and, thus, serve to extend the bounds of miscibility in polymer blends. The resulting concentration dependence can be introduced into the interaction parameter by expressing X as a polynomial in the concentration. Since ϕ_i, ϕ_j and functions thereof are all interdependent, the particular measure of concentration is arbitrary. Using $\phi_i - \phi_j$ gives for the excess free energy [9]:

$$\Delta G_M/kTV = (V_i N_i)^{-1} \phi_i \ln \phi_i + (V_j N_j)^{-1} \phi_j \ln \phi_j$$
$$+ \phi_i \phi_j [X_0 + X_1(\phi_i - \phi_j) + X_2(\phi_i - \phi_j)^2 + X_3(\phi_i - \phi_j)^3] \tag{6}$$

If all odd powers are omitted from Equation (6):

$$\Delta G_M/kTV = (V_i N_i)^{-1} \phi_i \ln \phi_i + (V_j N_j)^{-1} \phi_j \ln \phi_j + \phi_i \phi_j [X_0 + X_2(\phi_i - \phi_j)^2] \tag{7}$$

Equation (7) describes the symmetrical model of mixtures [9] according to which the mixing free energy is invariant to changes in composition that are symmetrical. The spinodal equation for the symmetrical mixture is then:

$$X_0 + [5(\phi_i - \phi_j)^2 - 8\phi_i\phi_j]X_2 = [(V_iN_i\phi_i)^{-1} + (V_jN_j\phi_j)^{-1}]/2 \qquad (8)$$

where

$$X_{sp} = V_r[X_0 + X_2(\phi_i - \phi_j)^2] \qquad (9)$$

The critical composition for a symmetrical mixture, unlike that of a simple Flory-Huggins blend [Equation (4)], is dependent on the magnitude of the interaction parameter.

The magnitude of the van der Waals interactions can be described by a series whose leading term corresponds to the well-known London equation for the dispersion energy [10]:

$$E_{ij} = -(3/4) I_{ij}\alpha_i\alpha_j \, r^{-6} \qquad (10)$$

where α_i is the polarizability of the ith molecule or chain unit separated by r from the jth unit, and I_{ij} is approximately equal to the ionization potential of the species. From Equation (10) it is seen that the mixing enthalpy associated with dispersive interactions always favors phase segregation. This result means that the interaction parameter for a blend is positive in the absence of chemical reaction between the components, reflecting the greater attraction between like segments than between unlike segments. Miscibility in such a system is due only to the combinatory entropy; as a result phase separation can be induced by increases in the molar mass of the constituents. This increase provides a convenient means to characterize the excess free energy of the mixture from determination of the critical values of molecular weight required to cause phase separation. The magnitude of the interaction parameter for the blend can be deduced from application of expressions such as Equations (3) or (8).

For macromolecules the combinatory entropy makes such a small contribution to the free energy that miscibility is usually restricted to those mixtures in which the components chemically interact and thereby effect a negative excess enthalpy. The diminution of such specific interactions from increasing thermal agitation may then occasion the observation of a lower critical solution temperature (LCST) in this type of mixture [11–15]. In the less common case of miscibility absent specific interactions, upper critical solution temperatures (UCST) are expected since the contribution of the combinatory entropy to the free energy of mixing is proportional to temperature.

The mixing theories on which Equations (1) and (6) are based describe "regular" solutions; that is, they assume the blend volume is the sum of the pure component volumes. Often, however, a difference in liquid structure (free volume) of the components gives rise to an excess volume with concomitant influence on the mix enthalpy and entropy [2,16,17]. The magnitude of these equation of state effects can be gauged from the dissimilarity in the thermal expansion coefficients of the component polymers. Components of different size (loosely speaking—chain molecules of different diameters) will usually form mixtures having a negative excess volume (a net contraction). Even when the latter is zero, however, the size differences can still influence the mixing free energy. Contraction of the mixture will make a negative contribution to both the enthalpy and entropy with the net effect invariably favoring demixing. Since this negative excess entropy is enhanced by increases in temperature, mixtures of polymers can exhibit LCST even in the absence of specific interactions. The competing effects of the combinatory entropy and of differences in liquid structure can in principle give rise to both upper and lower critical solution temperatures.

The present report is concerned with various mixtures involving polybutadiene (both the 1,2- and 1,4- isomers) and *cis*-1,4-polyisoprene. Whereas in all cases associated with positive excess enthalpies reflecting an absence of specific interactions [18–21], these mixtures exhibit marked differences in their phase behavior.

EXPERIMENTAL

The polydienes were either obtained from commercial sources or synthesized in house. All had narrow molecular weight distributions. The BR was either atactic 1,2-polybutadiene, 1,4-polybutadiene, or random copolymers thereof. The IR was 1,4-polyisoprene of high *cis* content.

A Perkin-Elmer DSC-2 was used for calorimetry. Heat capacity changes were recorded while heating 5–7 mg samples at 20° per minute from −125°C to 30°C. Annealing experiments consisted of heating at 320° per minute to the desired temperature followed by quenching at >200° per minute after varying dwell times. This cooling rate was sufficiently rapid that any changes in the phase structure during quenching were not observable. Crystalline samples were prepared by placing preweighed, dried samples contained in DSC specimen holders into small desiccators which were then maintained at −9°C for varying duration. Immediately prior to thermal measurements, these samples were placed in a small container of liquid nitrogen, transferred to the dry box enclosure and loaded into the DSC. The head was held at −13°C during this loading. The temperature was then reduced to −93°C, followed by heating through the melting point of the IR up to 60°C. The temperature was subsequently brought back to −93°C, followed by reheating through the glass transition. In all cases temperature changes were executed at 20° per minute.

RESULTS

Phase Behavior of Blends

The temperature of the glass to liquid transition was measured for various blends of polybutadiene and polyisoprene. In all cases the components were present at their critical concentration as given by Equation (5). The degrees of polymerization of the respective components required to effect distinct glass transitions for each component, providing evidence of heterogeneous phase morphology, were determined. Under certain circumstances the calorimetrically measured transitions were very broad; however, thermodynamic miscibility of these particular blends was corroborated by proton and carbon NMR measurements [22] and by observation of spontaneous interdiffusion of the separated pure components [18,20].

The interaction parameters determined for the blends are displayed in Table 1. These results were obtained from determination of the miscible compositions with the lowest critical interaction parameter [as calculated from Equation (3)] and the immiscible blend with the highest X_{cr}. It is observed that as the concentration of 1,2- units in the BR increases, there is a large increase in miscibility with IR. At high levels of 1,2- units, phase separation cannot be induced even at extremely high molecular weights which, given the absence of specific interactions, indicates a remarkable degree of miscibility. This miscibility suggests a near equivalence in polarizability between the respective chain units of the 1,2-polybutadiene and the 1,4-polyisoprene, along with a close similarity in liquid structure (or their degree of expansivity). At higher 1,4 content a larger mixing endotherm restricts miscibility to low molecular weights. Also displayed in Table 1 are the values of the interaction parameter calculated for the blends assuming the thermodynamics were described by the symmetrical model. From the determination of the miscible composition of the highest molecular weight components and the immiscible composition of the lowest molecular weight components, an interaction parameter was estimated using Equations (8) and (9). For this purpose it was assumed that X_2 was 20% of X_0 [6,23]. The Flory-Huggins theory underestimates the extent of the miscible region of the phase diagram; consequently, calculations of the interaction parameter for a blend based on it are seen to be underestimated.

The relative contributions to the mixing enthalpy for these blends from changes in van der Waals energy and from equation of state effects are not readily

Table 1. Results for mixtures of IR with BR copolymers.

Component[a]	Component	$X_{sp} \times 10^3$	
		Equation (3)	Equation (8)
92% 1,4-BR	IR	2.4	2.4
74% 1,4-BR	IR	1.7	1.9
59% 1,4-BR	IR	0.7	0.8
97% 1,2-BR	IR	<0.17[b]	<0.19[b]
97% 1,2-BR	92% 1,4-BR	2.3	3.7

[a]All blends contained in the Flory-Huggins critical concentration of components as calculated by Equation (5).
[b]Phase separation not observed.

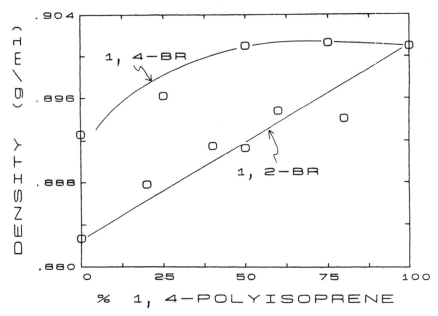

FIGURE 1. The density measured at room temperature for IR mixed with 86% 1,2-BR and 92% 1,4-BR. The latter blends exhibit deviation from additivity, indicative of a mismatch in the liquid structure of the pure components.

apparent. From inspection of solubility parameters for the polymers, however, it is clear that the exchange enthalpy between IR and BR becomes more endothermic (reduced miscibility) as the concentration of 1,4- units in the latter increases [24]. A closer similarity in polarizability between IR and 1,2-BR segments thus underlies in part the high degree of miscibility of their blends.

There will also be a contribution to the free energy of mixing from any differences in the liquid structure of the two components [2,16,17]. Displayed in Figure 1 is the density measured for IR blended with 1,4-BR and 1,2-BR, respectively. While the volumes of the latter mixtures are simply additive in the volume of the components, 1,4-BR/IR blends exhibit a negative excess volume. The blend densities reflect the differing contribution of equation of state effects to the liquid structure of the mixtures.

Although absence of an excess volume upon mixing does not imply non-additivity of the enthalpy and non-combinatory entropy, any such equation of state contributions to the free energy must be vanishingly small in blends of 1,2-BR with IR in order that miscibility be observed at high component molecular weights. Significant differences in liquid structure are made apparent by comparison of the thermal expansion coefficients of the components. The thermal expansion coefficient of 1,2-BR and the IR have been shown to be very nearly equivalent, indicating the similarity in their liquid structure [20]. The expansivities depend strongly on polybutadiene microstructure, however, resulting in a distinct mismatch in liquid structure between IR and BR of high 1,4- content [21]. These equation of state effects are responsible at least in part for the reduction in miscibility of IR with 1,4-BR.

A further consequence of significant equation of state effects is the potential for lower critical solution temperatures [21]. It is expected that when the temperature of a miscible blend of BR and IR is reduced, phase segregation will eventually transpire since the driving force for miscibility is the combinatory entropy. Such upper critical solution temperatures, however, are evidently below the glass transition temperatures of the mixtures herein and thus unobservable. In contrast, phase separation of miscible mixtures of IR with BR of high 1,4- content were induced by increases in temperature. Representative calorimetry results demonstrating this LCST are displayed in Figure 2 for a blend of 59% 1,4-BR and IR and in Figure 3 for a 92% 1,4-BR with IR. In contrast to these data, when the BR microstructure is less than 15% 1,4, no phase separation in blends with IR has ever been observed [18,20]. The absence of an LCST is consistent with increased similarity in the expansivities of the components at higher 1,2- microstructure.

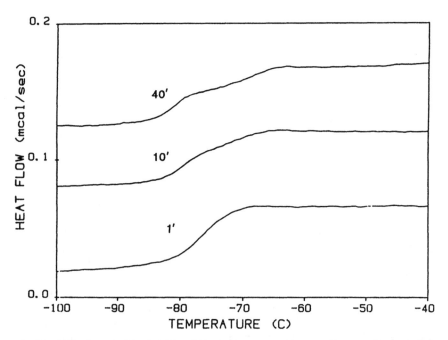

FIGURE 2. The glass transition behavior of blends of IR (N = 4500) with 59% 1,4-BR (N = 1700) at the critical concentration of components. The samples were heated from 30°C to 50°C and held at the latter temperature for the indicated time period, followed by quenching to −125°C. The displayed curves, corresponding to measurements made in the ensuing reheat, reflect the increasing extent of phase separation transpiring just above the LCST.

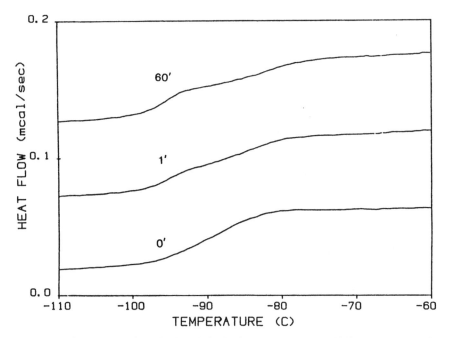

FIGURE 3. The glass transition behavior of blends of IR (N = 1940) with 92% 1,4-BR (N = 440) at the critical concentration. The samples were heated from 30°C to 75°C and held at the latter temperature for the indicated time period, followed by quenching to −125°C. The displayed curves, corresponding to measurements made in the ensuing reheat, reflect the increasing extent of phase separation transpiring just above the LCST.

Crystallization

The crystallization and melting behavior of a polymer can be altered by its presence in a miscible blend. The thermodynamic stability conferred upon the liquid state by formation of a miscible blend reduces the relative stability of the crystalline state and thus will lower the equilibrium melting point. By assuming equilibrium crystallization, the melting point depression can be related to the Flory interaction parameter [25,26]. For purely dispersive interactions between components, the largest change in melting temperature corresponds to ideal (athermal) mixing. For high polymers the resulting suppression of the melting point will be immeasurably small [18]. It is, therefore, expected that pure IR and IR in blends with 1,2-BR will exhibit equivalent melting behavior.

A series of IR/1,2-BR mixtures were isothermally crystallized at −9°C for varying duration up to 37 days; representative DSC scans of these mixtures are displayed in Figure 4. The measured melting temperatures, averaged over several samples, are tabulated in Table 2; they are equivalent within the precision of the data. These results are consistent with an interaction parameter of negligible magnitude in these blends.

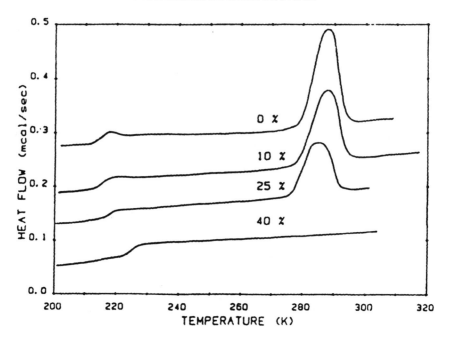

FIGURE 4. Representative DSC scans after crystallization for blends of 1,2-BR with the indicated volume fraction of BR. The temperature at which the heat capacity changed abruptly was in all samples shifted by about 2° in the ensuing scan after melting.

The pure IR attained about a 30% extent of crystallinity based on the measured heat of fusion. In blends the IR was significantly less crystalline, even after 37 days at −9°C. Any difference in IR crystallinity, however, between samples with 10% and 25% 1,2-BR was less than the precision in the measurement of the heat of fusion. It is interesting that for samples with greater than 25% 1,2-BR, no melting endotherms could be detected, even though the instrumental sensitivity was more than sufficient if the reduction in crystallinity of the IR was simply proportional to the concentration of 1,2-BR (Figure 4). It is generally expected that,

Table 2. Crystallization results for IR/1,2-BR blends.

% IR	T_{melt} (°C)	ΔH^a	Crystallinity[b]
100	4.6 ± 0.5	4.5 ± .3	29%
90	4.8 ± 1.2	2.7 ± .9	18%
75	4.3 ± 2.3	2.6 ± .7	17%
60	—	0	0%

[a]Calories per gram of IR after 37 days at −9°C.
[b]Based on a perfect heat of fusion of 15.3 cal per gram [36].

while miscible blending can alter the rate of crystallization, the ultimate degree of crystallinity will be unchanged [27]. Since 37 days does not necessarily correspond to equilibrium, the lower crystallization levels of the IR in the blends does not necessarily suggest that the crystallization process has been suppressed.

In both the pure IR and the blends, the glass transition temperature was observed to decrease about 2° as a result of melting. This influence of crystallinity on T_g has been previously reported, with a slightly higher T_g of semicrystalline IR attributed to reduced mobility of the amorphous chains adjacent to crystallites [28].

CONCLUSIONS

The degree of miscibility of the various IR/BR mixtures is not obvious from inspection of the chemical structures. IR and 1,2-BR, despite their distinct differences in chemical structure, form blends that are always homogeneous. On the other hand, the polarizability and liquid structural differences between IR and 1,4-BR limit their miscibility to low molecular weights. As the 1,4- content of BR increases, the thermodynamic behavior of BR/IR blends ranges from nearly ideal mixing to the occurrence of a LCST due to negative excess mixing volumes. Whereas the observation of lower critical solution temperatures in polymer mixtures in which no specific interactions exist is unusual, the same mismatch in liquid structure gives rise to the LCST commonly observed in polymer solutions.

The melting point of the polyisoprene was unaltered by blending, in accordance with a very low magnitude of X. The crystallization rate was found to be suppressed in blends. This rate is not unexpected, since crystallite growth requires the 1,2-BR chains to disentangle and diffuse away from the growing spherulites.

REFERENCES

1. Flory, P. J. *Principles of Polymer Chemistry.* Ithaca, NY:Cornell (1953).
2. Mandelkern, L., J. E. Mark, U. W. Suter and D. Y. Yoon (eds.). *Selected Works of Paul J. Flory, Vol. I.* Stanford Press, p. 409 (1985).
3. de Gennes, P. G. *Scaling Concepts in Polymer Physics.* Ithaca, NY:Cornell (1979).
4. Stein, R. S. and G. Hadziioannou. *Macromolecules,* 17:1059 (1984).
5. Murray, C. T., J. W. Gilmer and R. S. Stein. *Macromolecules,* 18:996 (1985).
6. Roland, C. M. and C. A. Trask. *Polymer Bull.* (in press).
7. Cifra, P., F. E. Karasz and W. J. MacKnight. *Macromolecules,* 21:446 (1988).
8. Curro, J. G. and K. S. Schweizer. *J. Chem. Phys.,* 88:7242 (1988).
9. Guggenheim, E. A. *Applications of Statistical Mechanics.* Oxford:Clarendon (1966).
10. Hirschfelder, J. O., C. F. Curtiss and R. B. Bird. *Molecular Theory of Gases and Liquids.* New York:Wiley (1954).
11. Alexandrovich, F. E., F. E. Karasz and W. J. MacKnight. *Polymer,* 18:1022 (1977).
12. Shayama, M., H. Yang, R. S. Stein and C. C. Han. *Macromolecules,* 18:2179 (1985).
13. Maconnachie, A., R. P. Kambour, M. W. Dwain, S. Rostami and D. J. Walsh. *Macromolecules,* 17:2645 (1984).
14. Goh, S. H., D. R. Paul and J. W. Barlow. *Polym. Eng. Sci.,* 22:34 (1982).

15. Sham, C. K. and D. J. Walsh. *Polymer,* 28:804 (1987).

16. Patterson, D. *Rub. Chem. Tech.,* 40:1 (1967).

17. Patterson, D. *Poly. Eng. Sci.,* 22:64 (1982).

18. Roland, C. M. *Macromolecules,* 20:2557 (1987).

19. Roland, C. M. *J. Polym. Sci. Polym. Phys. Ed.,* 26:839 (1988).

20. Trask, C. A. and C. M. Roland. *Macromolecules* (in press).

21. Trask, C. A. and C. M. Roland. *Polym. Comm.* (in press).

22. Roland, C. M., C. A. Trask, K. McGrath and J. Miller (to be published).

23. Singh, R. R. and W. A. Van Hook. *Macromolecules.* 20:1855 (1987).

24. Roland, C. M. and C. A. Trask. 134th Natl. Meeting Rubber Div., Amer. Chem. Soc., Cincinnati, Ohio (Oct. 1988).

25. Nishi, T. and T. T. Wang. *Macromolecules,* 8:909 (1975).

26. Rim, P. B. and J. P. Runt. *Macromolecules,* 17:1520 (1984).

27. Runt, J. P. and L. M. Martynowicz. *Adv. Chem. Ser.,* 211:111 (1985).

28. Burfield, D. R. and K-L. Kim. *Macromolecules,* 16:1170 (1983).

10

Polyurethane-Polyvinyl Chloride Interpenetrating Polymer Networks

M. OMOTO,* D. KLEMPNER* and K. C. FRISCH*

INTRODUCTION

THE CHEMICAL AND physical combination of two or more structurally dissimilar polymers has been of commercial and academic interest for a number of years since it provides a convenient route for the modification of properties to meet specific needs. It has been used to facilitate processing and to impart flexibility, tensile and impact strength, chemical resistance, weatherability, flammability resistance, and a variety of other properties [1–4]. The physical properties of the combined polymers depend not only on the properties of the constituent polymers but also on the way they are combined.

Mechanical blending is the oldest way of physically combining two or more linear polymers and is accomplished by mixing the polymers in a liquid state, e.g., melt, solution, or dispersion.

Chemical combinations of polymers can be defined as the combination of two or more types of polymers (or corresponding monomers) generally via covalent bonds and can be achieved by random, alternate, block, or graft copolymerization.

Interpenetrating polymer networks (IPNs) are a novel type of polymer blend composed of crosslinked polymers. They are more or less intimate mixtures of two or more distinct crosslinked networks with no covalent bonds between the polymers, i.e., polymer A crosslinks only with other molecules of polymer A, and polymer B crosslinks only with other molecules of polymer B.

Thus, IPNs may be described as combinations of chemically dissimilar polymers in which the chains of one are completely entangled with those of the other. The entanglements must be of a permanent nature and are made so by this homocrosslinking of the two polymers. IPNs can be of various chemical types

*Polymer Technologies, Inc., University of Detroit.

111

and can be synthesized in a variety of ways. There are two basic techniques for producing IPNs. In the first technique, a crosslinked polymer of type A is swollen with a second monomer of type B, plus crosslinking agents. This mixture is followed by the polymerization and crosslinking of polymer B *in situ*. The second technique consists of combining the linear polymers, prepolymers, or monomers of the two polymer types, together with their respective crosslinking agents, in some liquid form, e.g., bulk (melt), solution, or dispersion. This combination is followed by the simultaneous polymerization and crosslinking of the two polymers. Care must be taken in the selection of the polymers to prevent reaction occurring between them. It is also preferable that the polymers be of different chemical types so that the resulting material will be more than just a copolymer. Interest centers in these materials for a variety of reasons.

1. IPNs represent a mode of blending two or more polymers to produce a mixture in which phase separation is not so extensive as would be otherwise. In fact, it is the only way of combining crosslinked polymers. Normal blending or mixing of polymers results in a multi-phase morphology due to the well-known thermodynamic immiscibility of polymers. This immiscibility is due to the relatively small gain in entropy upon mixing the polymers due to contiguity restrictions imposed by their large chain length [5]. However, if mixing is accomplished on a lower molecular weight level and then polymerization is accomplished simultaneously with crosslinking, phase separation may be kinetically controlled since the entanglements will have been made permanent by the crosslinking. In other words, phase separation cannot occur without breaking covalent bonds.

IPNs synthesized to date exhibit varying degrees of phase separation, dependent principally on the miscibility of the polymers. With highly incompatible polymers, the thermodynamics of phase separation are so powerful that it occurs substantially before the kinetic ramifications (crosslinking) can prevent it. In these cases only small gains in phase mixing occur. In cases where the polymers are more compatible, phase separation can be almost completely circumvented. Note that complete miscibility (an almost impossible situation) is not necessary to achieve complete phase mixing, i.e., interpenetration, since the "permanent" entanglements produced by interpenetration prevent phase separation. With intermediate situations of miscibility, intermediate and complex phase behavior (morphology) results. Thus, IPNs have been reported with dispersed phase domains ranging from a few microns (the largest) [6], to a few hundred angstroms (intermediate) [7], to those with no resolvable domain structure (complete mixing) [8].

2. IPNs represent a special example of topological isomerism [9] in macromolecules [10]. Some permanent entanglements between the different crosslinked networks are inevitable in any sufficiently intimate mixture of the crosslinked networks. These represent examples of catenation in polymer systems, i.e., different ways of imbedding these molecules in three-dimensional space. Permanent entanglements are hindering constraints on the motion of segments and ought to simulate covalently bound chemical crosslinks [11]. Simplified theoretical models of such permanent entanglements [12] exhibit a surprisingly large nonlinear elastic restoring force unlike that expected with chemical crosslinks from ideal rubber elasticity theory.

3. The combining of varied chemical types of polymeric networks in different compositions, often resulting in controlled different morphologies, has produced IPNs with synergistic behavior. For example, if one polymer is a glass (glass transition, T_g, above room temperature) and the other is elastomeric (T_g below room temperature), one obtains a reinforced rubber, if the elastomer phase is the continuous, predominant one, or a high impact plastic if the glassy phase is continuous [13]. In the case of more complete phase mixing, enhancement in numerous mechanical properties is due to the increased physical crosslink density due to this interpenetration. IPNs have been synthesized with intermediate maxima versus network composition in bulk properties such as tensile strength [14–22], impact strength [16,17], and thermal resistance [16–18,20] as well as intermediate minima in surface properties such as critical surface tension. Certain of these synergistic properties of IPNs, such as the mechanical properties, reflect their special architecture. Others, such as thermal resistance, may simply be consequences of the fact that IPNs are usually blends of different chemical types of networks and may have little to do with their special structure [23].

It is the "intermediate" situation of miscibility (referred to earlier) that is the focus of attention in the present study. Normally, single phase materials (i.e., simple homopolymers and copolymers and completely miscible polymer blends) have a relatively narrow glass transition range (20–30°C). Incompatible polymer blends exhibit two such transitions, corresponding to the T_gs of the component polymers [24,25]. On the other hand, semimiscible IPNs (the "intermediate" situation) made of polymers with high and low glass transition temperatures can lead to a very broad transition range [26,27].

In this study, it was desired to prepare mechanical and acoustical energy absorbing IPNs. Polymeric systems with broad T_gs (over a wide temperature and frequency range) would be most effective in this application because when polymers are at their glass transition, the time required to complete an average coordinated movement of the chain segments approximates the length of time of the measurement. If dynamic or cyclical mechanical motions are involved, such as vibrational or acoustic energy, the time required to complete one cycle, or its inverse—the frequency—becomes the time unit of interest. At the glass transition conditions, which involve both temperature and frequency effects, the conversion or degradation of mechanical or acoustical energy to heat reaches its maximum value. Thus, polymeric systems with damping behavior which span a broad temperature range (and therefore frequency range) centered approximately around room temperature will be the most effective energy absorbers, i.e., they will possess a high tan δ spanning a broad temperature range.

In the present study, pseudo interpenetrating polymer networks (pseudo IPNs—only 1 polymer crosslinked) of linear polyvinylchloride and crosslinked polyurethane were investigated as a system which might possess a broad transition. The transitions were determined by dynamic mechanical spectroscopy, with the tan δ behavior being the main indication of energy absorption capabilities, since it is well-known that the loss factor is closely related to the damping properties of these systems [14,28,29].

Blends of polyurethanes and poly(vinyl chloride) have been studied a great deal recently and show promising results in practical and scientific aspects. Of major concern is whether these systems are miscible, what factors influence their miscibility, and even what defines true miscibility. PU/PVC polyblends offer increased flexibility, tensile strength, impact strength, fire retardancy [30], and acoustic damping [31]. They can be used as foams, elastomers, coatings, adhesives, and plastics.

In a recent study, it was shown that blends of chlorinated poly(vinyl chloride) (PVC) and a polyester based polyurethane (PU) containing poly(tetramethylene adipate) capped with 2-hydroxyethylacrylate were partially miscible over the entire composition range at 200–220°C as determined by differential scanning calorimetry (DSC), transmission electron microscopy (TEM), and dynamic mechanical analysis (DMA) [32]. Garcia attributed the miscibility to strong hydrogen bonding between the α-hydrogen of CPVC and the carboxyl of poly(tetramethylene adipate). Less importantly was the interaction of the α-H and the carboxyl of the urethane bond. Similarly, it was shown that in a PVC/PU blend, where the PU consisted of 4,4′-diphenylmethane diisocyanate (MDI) plus a mixture in different ratios of polycaprolactone (PCL) and either polytetramethylene oxide (PTM) or polyethylene oxide (PEO), miscibility increased with increasing PCL chain length. However, the higher molecular weight caused the PU to slowly crystallize out at blend ratios greater than 50:50 by weight. Also, miscibility was affected by the type of polyether segment and its molecular weight while the mechanical and viscoelastic properties depended on PCL content. The authors also state that the temperature dependent partial miscibility broadens the glass transition and imparts high damping over a wide range of temperatures.

In a similar study, it was reported that PCL based polyurethanes were miscible with PVC at all compositions over a broad, accessible temperature range [33]. This miscibility is most likely due to the hydrogen bonding between the ester group in PCL and the $H-C-Cl$ in PVC. On a broader basis, there is evidence to suggest that polyether based urethanes are more miscible with PVC than similar polyester based polyurethanes [31]. This suggestion is based on thermal analysis which yielded only a single T_g. PVC interaction with ether sequences was ruled the cause.

However, Wang and Cooper utilized DSC, dynamic mechanical analysis, stress–strain, infrared peak position, and infrared dichroism data to show that THF solution prepared polyether based PU/PVC blends were partially miscible whereas THF + dioxane solution prepared blends were not miscible [34]. It should be noted that PU is soluble in THF and/or dioxane but PVC is soluble in THF or the mixture, but not dioxane alone. Another conclusion drawn was that whereas "true" miscibility was not achieved in all films, mechanical compatibility was, i.e., there was a plasticization effect on PVC in all films. Again, in another study, a PTMO + TDI poly(ether urethane)-PVC blend was shown to be essentially heterogeneous [30], as evidenced by morphological and thermodynamical data. Kalfoglou suggests [30] that the useful properties obtained by blending are due to good adhesion between the phases. Further, it is suggested that the shifts in transitions with composition changes in PU are due to good adhesion between the phases. Further, it is suggested that the shifts in transitions with composition changes in PU are due to adhesion caused by hardening of the segmental mobility and for PVC are due to a plasticization effect [30]. The theory of adhesion between the phases is also used by Blaga [35] in his study of PU blends. Blaga et al. [35] found that their polyether diol + MDI polyurethane-PVC blend was heterogeneous in ratios of 20:1 to 5:1 and that PVC is present in the PU matrix as dispersed particles. The good properties again are attributed to adhesion between the phases. Piglowski et al. also attribute partial miscibility to adhesion by suggesting the presence of a semi-compatible intermediate layer at the phase boundary [36]. Other studies have been carried out to determine the effect of the components of PU [37] and show either partial miscibility or immiscibility [38].

Thus, in the present study, pseudo IPNs of PU and PVC were prepared for the first time and characterized by DMA. The PUs contained varying ratios of PCL polyol and polypropylene glycol (PPG), in order to obtain IPNs of varying morphology.

In earlier studies in our laboratory [39], pseudo IPNs composed of 100% PCL-based PU and poly(vinylchloride) were prepared and their phase structure studied by dynamic mechanical spectroscopy. A single tan δ peak occurred at a temperature between the glass transition temperature of polyurethane and PVC. This data indicates that these pseudo IPNs have a single-phase nature at room temperature. To obtain the lower solution temperature (LCST) of these pseudo IPNs (if any), the samples were annealed at 200°C, quenched in liquid nitrogen, and then dynamic mechanical spectra were measured. However, the spectra did not show any two-phase morphology (no change). Annealing above 200°C could not be carried out because of the decomposition of polyurethane and PVC. Therefore, it was felt that the miscibility will not decrease as temperature is raised, in agreement with Ball and Galyer [32]. Polyether glycols, such as polyoxypropylene glycol (PPG)- based polyurethanes are known to be immiscible with PVC [39].

If the mixed polyols of PCL and PPG are used to prepare polyurethanes, the miscibility of the polyurethanes with PVC will be decreased and an LCST behavior might occur.

Earlier studies showed that pseudo IPNs composed of 80% PCL/20% PPG-based polyurethane and PVC exhibited homogeneous behavior at room temperature [39].

In this study, pseudo IPNs composed of varying ratios of PCL to PPG-based polyurethane and PVC were prepared and their morphology determined by DMA. The 80/20 PCL/PPG-based IPNs were exposed to thermal treatments in order to obtain a phase diagram, if an LCST occurred.

EXPERIMENTAL

Materials

The materials used in this study are listed in Table 1. The polyols, crosslinking agent, and PVC were dried by heating at 70–80°C under vacuum overnight. Tetrahydrofuran (THF) was dried over molecular sieves (Type 4A). 4,4′-

Table 1. Materials.

Chemical Composition	Tradename	Supplier	Mol. Wt.	Eq. Wt.	Designation
Poly(oxypropylene glycol)	Pluracol P-2010	BASF	1,992	996	PPG-2000
Poly(ϵ-caprolactone) glycol	Tone 0240	Union Carbide	1,996	998	PCL-2000
4,4′-Diphenylmethane diisocyanate	Mondur M	Mobay	250	125	MDI
Triethanolamine		Matheson Co. Inc.	149	49.7	TEA
Polyvinylchloride	VC-106PM	Borden Chem.	58,000		PVC
Dibutyltin dilaurate	T-12	M & T Chem.			T-12
Tetrahydrofuran		Fisher Scientific			THF
Cobalt salt	BC-103	Interstab Chem.			BC-103

Diphenylmethane diisocyanate (MDI), dibutyltin dilaurate (T-12 catalyst) and cobalt salt (BC-103, PVC stabilizer) were used as received.

Preparation of Prepolymer

The following procedure was used for the preparation of isocyanate-terminated polyurethane prepolymers based on 80% polycaprolactone glycol (PCL)/20% polyoxypropylene glycol (PPG). A resin kettle with nitrogen inlet, stirrer, and thermometer was charged with 3 moles of MDI. The MDI was melted at 50–60°C and 0.4 moles of PPG were added with stirring. The reaction was carried out at 70–80°C under a dry nitrogen atmosphere until the isocyanate content reached the theoretical value. Then 1.6 moles of PCL were slowly added with stirring. The reaction was carried out at 70–80°C until the isocyanate content reached the theoretical value. The isocyanate content was determined by the di-*n*-butylamine titration method. The prepolymer was dissolved in THF (concentration 50%). Table 2 lists the molar ratios of prepolymer used for this study.

Preparation of Pseudo-IPN

The polyurethane prepolymer solution in THF was mixed in the desired ratio with a PVC solution in THF (concentration 10%). Both components were mixed well in a closed Erlenmeyer flask. The crosslinking agent (triethanolamine) was then added. The amount of crosslinking agent was calculated from the isocyanate content of the prepolymer so that the isocyanate index was 105 (NCO/OH = 1.05). PVC stabilizer (BC-103, cobalt salt) was added to the mixed solution (11% of the weight of PVC). After these components were mixed well, catalyst (T-12) was added (0.5% of the weight of polyurethane prepolymer). Films of this solution were cast on polypropylene sheets using a doctor blade and kept in a desiccator for about 30 minutes to avoid the effects of moisture. The films were then placed in an oven and cured for 16 hours at 80°C.

Annealing and Quenching of Samples

A sample was cut from an IPN film in the same form (rectangle) as the specimen for dynamic mechanical analysis. This cut sample was sandwiched between polytetrafluoroethylene films to release easily and covered with aluminum foil to keep its original shape. The sample was put in a test tube filled with nitrogen gas. The test tube with the sample was kept in an oil bath controlled at the desired

Table 2. Polyurethane prepolymer formulation.

Prepolymer Code	Molar Ratio		
	MDI	PCL-2000	PPG-2000
A	3	2	—
B	3	1.6	0.4

Table 3. Composition and turbidity of pseudo IPNs.

Sample Code	Prepolymer Composition	PU Wt. %	PVC Wt. %	Turbidity
A-1	100% PCL	100	—	T
A-2	100% PCL	80	20	C
A-3	100% PCL	60	40	C
A-4	100% PCL	50	50	C
A-5	100% PCL	40	60	C
A-6	100% PCL	20	80	C
B-1	80% PCL/20% PPG	100	—	T
B-2	80% PCL/20% PPG	80	20	C
B-3	80% PCL/20% PPG	60	40	C
B-4	80% PCL/20% PPG	50	50	C
B-5	80% PCL/20% PPG	40	60	C
B-6	80% PCL/20% PPG	20	80	C
C	—	—	100	C

C = clear film; T = turbid.

temperature. After annealing, the sample was quenched in liquid nitrogen and the dynamic mechanical properties determined.

Dynamic Mechanical Spectroscopy

Measurements of the dynamic mechanical properties were carried out on a Rheovibron Dynamic Viscoelastomer DDV II (Toyo Manufacturing Co.) at a scanning rate of 1–2°C per minute in the glass transition region and every 3–5°C per minute in the non-transition region. The specimens were in the form of rectangular films with the dimensions of 3.5 cm × 0.4 cm × 0.01 cm. The measurements were carried out over a temperature range of −50°C to 120°C at a frequency of 110 Hz.

Tensile Tests

Tensile tests were carried out on an Instron Table Model 1130 Universal Tester. The procedures are described in ASTM D-1708. A microdie was used to cut the samples. The crosshead speed was chosen as two inches per minute.

RESULTS AND DISCUSSION

The composition and turbidity of the pseudo IPN films composed of 80% PCL/20% PPG-based polyurethane and PVC are shown in Table 3, which also contains pseudo-IPNs composed of 100% PCL-based polyurethane and PVC. Both polyurethane films were turbid but all IPN and PVC films were transparent.

Table 4. Dynamic mechanical data.

Sample Code	Temp. at E''_{max} °C (K)	Temp. at tan δ_{max} °C (K)	Height of tan δ Peak	½ Width tan δ Peak
A-1-a, Turbid sample	−33 (240), −2 (271)	−25 (248)	0.49	51
A-1-b, Clear sample after heating	−32 (241), 0 (273)	−22 (251)	1.575	17
A-2	−19 (254)	−6 (267)	1.05	25
A-3	−3 (270)	16 (289)	0.85	36
A-4	10 (283)	34 (307)	0.91	37
A-5	20 (293)	45 (318)	0.76	51
A-6	49 (322)	86 (359)	0.68	53
B-1-a, Turbid sample	−36 (237), −5 (268)	−29 (244)	0.665	16
B-1-b, Clear sample after heating	−39 (234), −2 (271)	−25 (248)	1.69	18
B-2	−20 (253)	−3 (270)	1.005	30
B-3	−9 (264)	18 (291)	0.68	52
B-4	4 (277)	39 (312)	0.675	54
B-5	13 (286)	63 (336)	0.522	86
B-6	50 (323)	94 (367)	0.610	50
C	77 (350)	98 (371)	0.97	37

Tensile Properties

Tensile strength and elongation at break were measured for all IPNs, polyurethane and PVC films. The results are shown in Table 4 and Figures 1 and 2. The tensile strength and elongation of pseudo IPNs show an S-shaped dependence on composition. The minimum in tensile strength at 80% polyurethane is attributed to an interruption in hydrogen bonding in the polyurethane. Strength increases at lower polyurethane content due to the increasing content of the PVC and interpenetrating effect (which manifests itself in terms of maxima at 20% polyurethane). The elongation essentially increases with polyurethane content, as expected.

Dynamic Mechanical Properties

Figures 3–10 show the dynamic mechanical spectra of the pseudo-IPNs. Figure 3 is the spectrum of the turbid sample and shows a small tan δ and multiple E'' peaks. This turbid sample became transparent on heating for a few minutes at 100°C. Figure 4 was obtained on this transparent sample and shows a very high tan δ peak with only one E'' peak. This difference may be explained by the crystallinity of PCL.

All the IPN spectra show a single E'' peak and a tan δ peak, indicating their single-phase nature. The results are summarized in Table 4 where the IPNs composed of 100% PCL-based polyurethane and PVC are also included. Comparing the IPNs of 100% PCL-based polyurethane/PVC and 80% PCL:20% PPG-based polyurethane/PVC, it can be seen that the tan δ peaks of the IPNs of 100% PCL-based polyurethane/PVC are higher and narrower than those of the IPNs of 80% PCL:20% PPG-based polyurethane/PVC. This data shows that the IPNs of 80% PCL:20% PPG-based polyurethane/PVC are less miscible than those of 100% PCL-based polyurethane/PVC, since a broader tan δ peak indicates increased heterogeneity. The tan δ peak is extremely broad with IPNs of 40% PU. This fact indicates their potential utility as energy absorbing materials.

Figures 11 and 12 show the behavior of the temperature at E''_{max} vs. composition. The plots give the usual functional dependence on composition, i.e., concave upwards towards the weight-average line. Figures 13 and 14 show the plots of the temperature at tan δ_{max} vs. composition. These plots show an S-shaped dependence on composition. There are many equations [5] which describe the glass transition temperatures of blends:

$$\frac{1}{T_g} = \frac{w_1}{T_{g1}} + \frac{w_2}{T_{g2}} \tag{1}$$

$$T_g = \frac{w_1 T_{g1} + k w_2 T_{g2}}{w_1 + k w_2} \tag{2}$$

$$T_g = \frac{w_1 T_{g1} + k w_2 T_{g2}}{w_1 + k w_2} + q w_1 w_2 \tag{3}$$

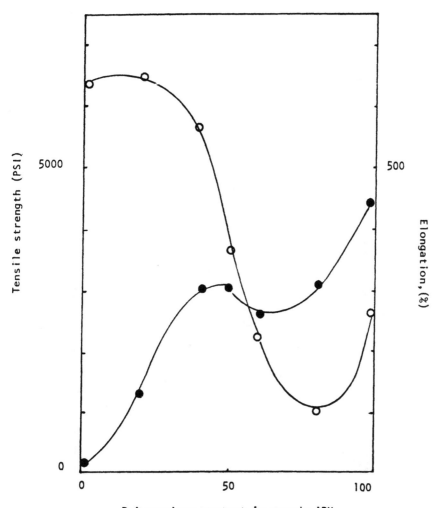

FIGURE 1. Pseudo IPN, PU (100% PCL)/PVC.

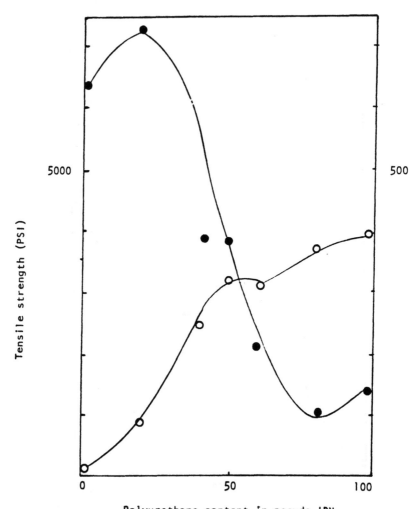

FIGURE 2. Pseudo IPN, PU(PCL/PPG: 80/20)/PVC.

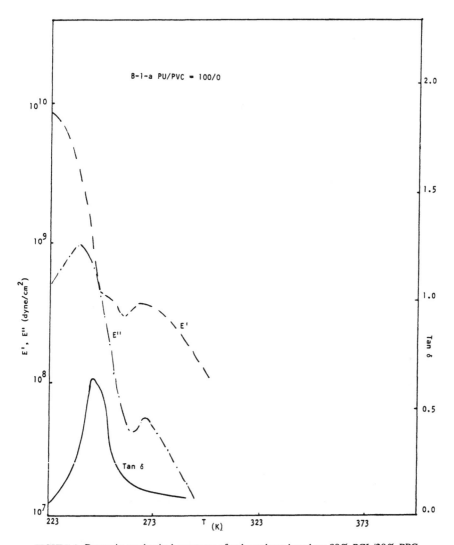

FIGURE 3. Dynamic mechanical spectrum of polyurethane based on 80% PCL/20% PPG.

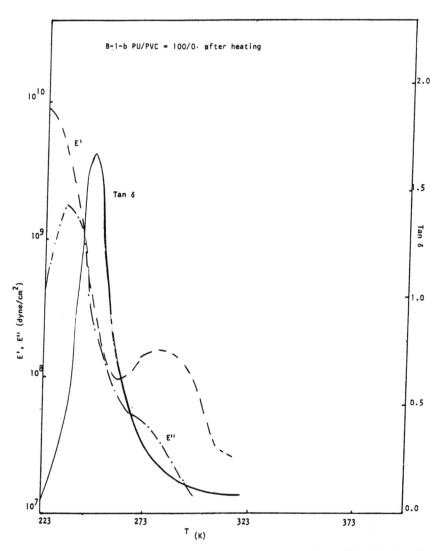

FIGURE 4. Dynamic mechanical spectrum of polyurethane based on 80% PCL/20% PPG, clear film after heating for a few minutes at 100°C.

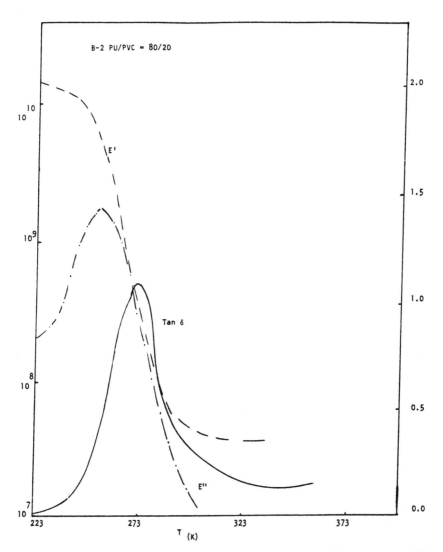

FIGURE 5. Dynamic mechanical spectrum of pseudo IPN composed of 80% polyurethane (80% PCL:20% PPG based) and 20% PVC.

125

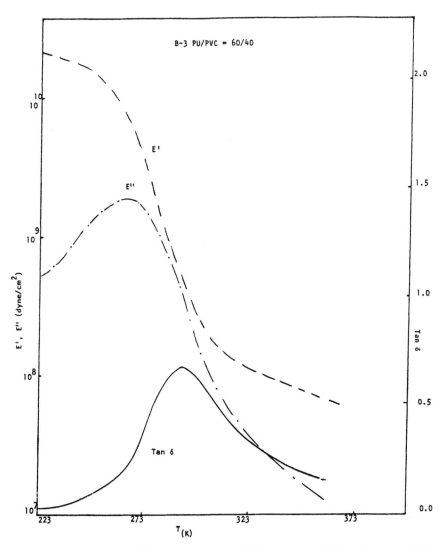

FIGURE 6. Dynamic mechanical spectrum of pseudo IPN composed of 60% polyurethane (80% PCL:20% PPG based) and 40% PVC.

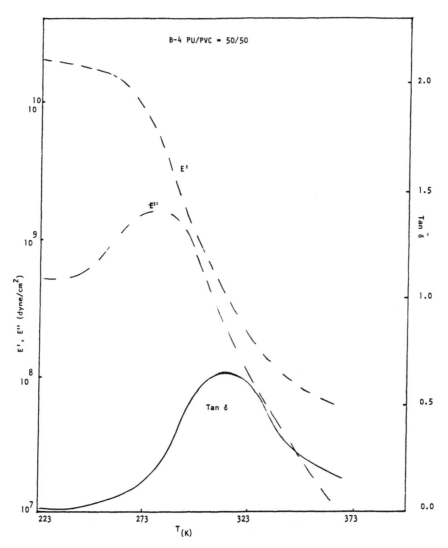

FIGURE 7. Dynamic mechanical spectrum of pseudo IPN composed of 50% polyurethane (80% PCL:20% PPG based) and 50% PVC.

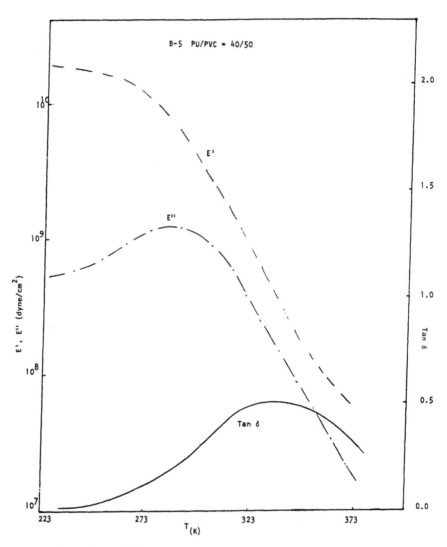

FIGURE 8. Dynamic mechanical spectrum of pseudo IPN composed of 40% polyurethane (80% PCL:20% PPG based) and 60% PVC.

128

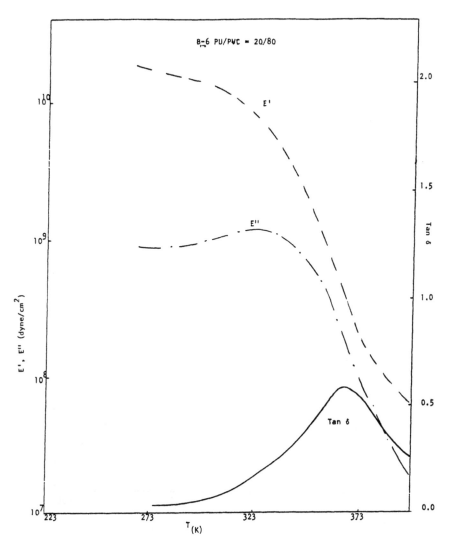

FIGURE 9. Dynamic mechanical spectrum of pseudo IPN composed of 20% polyurethane (80% PCL:20% PPG based) and 80% PVC.

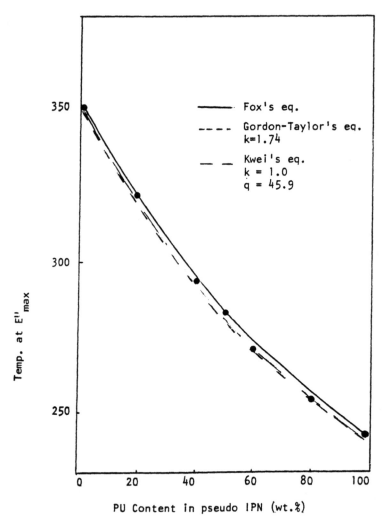

FIGURE 10. Dynamic mechanical spectrum of 100% PVC.

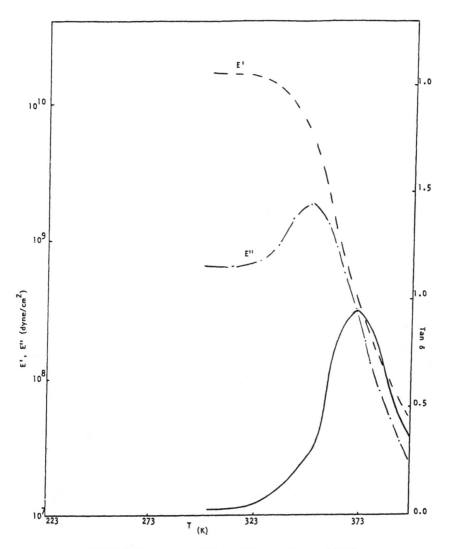

FIGURE 11. Temperature at E''_{max} vs. PU content in pseudo IPN.

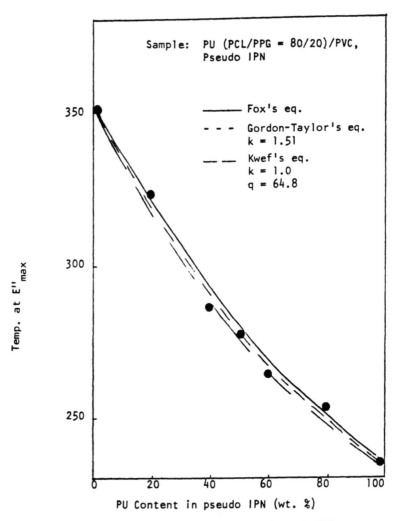

FIGURE 12. Temperature at E''_{max} vs. PU content in pseudo IPN.

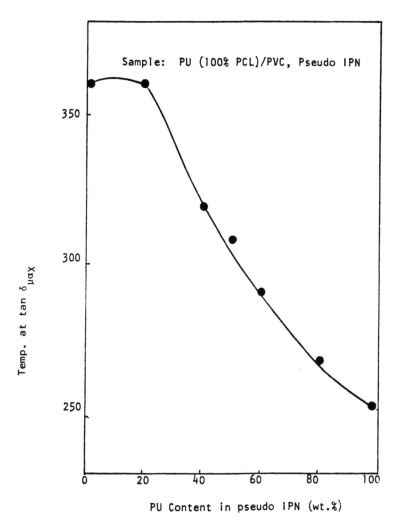

FIGURE 13. Temperature at tan δ_{max} vs. PU content in pseudo IPN.

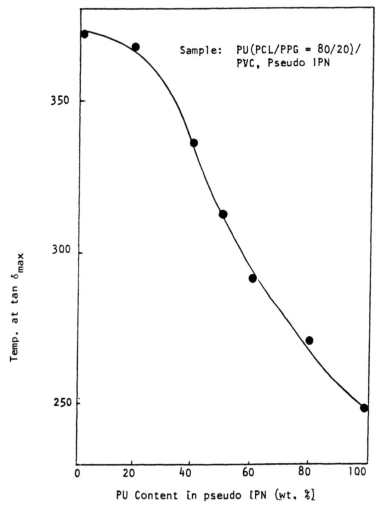

FIGURE 14. Temperature at tan δ_{max} vs. PU content in pseudo IPN.

where T_g, T_{g1} and T_{g2} are the glass transition temperatures of the blend, polymer 1 and 2, respectively, w_1 and w_2 are the weight fractions of polymer 1 and polymer 2 in the blend, and k and q are coefficients. The three curves in Figures 11 and 12 are drawn according to Equations (1–3). The k values in Equation (2) are 1.74 and 1.51 for the IPNs of 100% PCL-based polyurethane/PVC and 80% PCL 20% PPG-based polyurethane/PVC, respectively. The k values in Equation (3) are 1.0 for both IPNs and q values are -46 and -65 for the IPNs of 100% PCL-based polyurethane/PVC and 80% PCL:20% PPG-based polyurethane/PVC, respectively. The temperatures at E''_{max} were well predicted by these three equations.

Effect of Annealing on Pseudo-IPNs

IPN OF POLYURETHANE/PVC: 80/20

Figures 15 and 16 show the dynamic mechanical spectra obtained on samples annealed at 160°C and 200°C, respectively, followed by quenching. Only a single tan δ peak results in these spectra. These data indicate that the two-phase region of this sample may be above 200°C.

On the sample that was kept for 17 days at room temperature, a small tan δ peak at 237 K (-36°C) appeared, as shown in Figure 17. The main tan δ peak at 253 K (-20°C) diminished after storage for 28 days, as shown in Figure 18. These data are listed in Table 6.

IPN OF POLYURETHANE/PVC: 60/40

Two tan δ vs. T results on the samples annealed above 140°C are shown in Figures 19–23. The peak at the lower temperature is very small, whereas the peak at the higher temperature is large (main peak). The two peaks are shown more clearly on the E'' spectrum. The main tan δ peak (291–294 K) shifted to higher temperature (301–305 K) after annealing. The tan δ peak at lower temperature (240–245 K) is very close or a little lower than that obtained on the 100% polyurethane (244–248 K). These data indicate that the sample annealed above 140°C showed a two-phase morphology.

Dynamic mechanical spectra were measured again on the samples that were kept for 20 and 40 days at room temperature after annealing at 200°C and 220°C, respectively. As seen in Figures 24 and 25, the main tan δ peaks at 313 K (40°C) became broader than those obtained soon after annealing. These data are listed in Table 7.

IPN OF POLYURETHANE/PVC: 50/50

Two tan δ (and also two E'') peaks were obtained on the samples annealed above 180°C, as shown in Figures 27–29, whereas a single tan δ peak was obtained on the sample annealed at 160°C, as shown in Figure 6. The main tan δ peak (312 K) shifted to a higher temperature (316–317 K) after annealing. These data suggest that spinodal decomposition would occur between 160°C and 180°C.

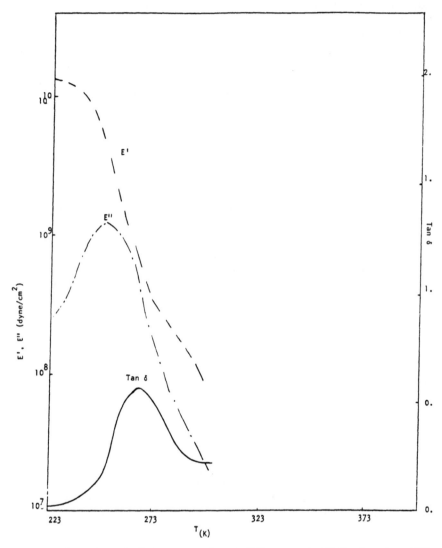

FIGURE 15. Dynamic mechanical spectrum of pseudo IPN composed of 80% polyurethane (80% PCL:20% PPG based) and 20% PVC, annealed at 160°C for 1.5 hours.

136

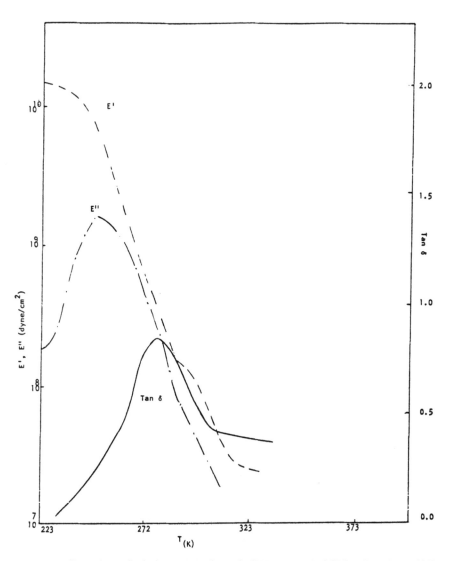

FIGURE 16. Dynamic mechanical spectrum of pseudo IPN composed of 80% polyurethane (80% PCL:20% PPG based) and 20% PVC, annealed at 200°C for 15 minutes.

137

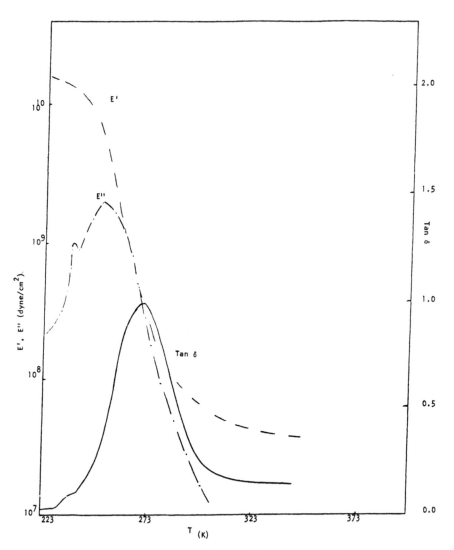

FIGURE 17. Dynamic mechanical spectrum of pseudo IPN composed of 80% polyurethane (80% PCL:20% PPG based) and 20% PVC, measured after 17 days.

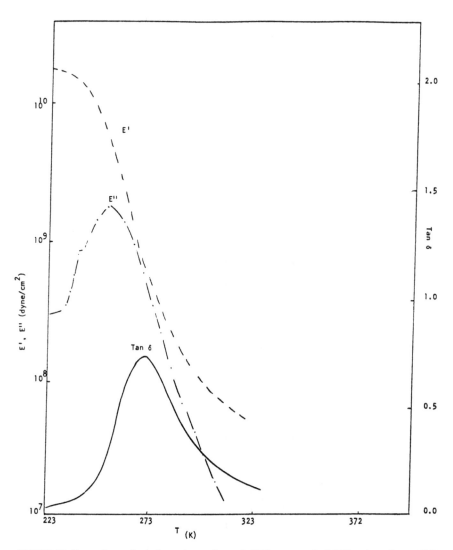

FIGURE 18. Dynamic mechanical spectrum of pseudo IPN composed of 80% polyurethane (80% PCL:20% PPG based) and 20% PVC, measured after 28 days.

139

Table 5. Dynamic mechanical data of IPNs before and after annealing
sample: B-2 [80% PU (80% PCL/20% PPG)/20% PVC].

Annealing Conditions	Days After Preparing Film	Temp. at E''_{max} (K)	Temp. at $\tan \delta_{max}$ (K)	Height of $\tan \delta$ Peak	½ Width of $\tan \delta$ Peak (K)	Temp. at Another Peak or Shoulder		Figure No.
						E'' (K)	$\tan \delta$ (K)	
Not annealed (I)	2	254	271	1.095	28			5
Not annealed (II)	17	253	270	1.005	30	237	237	15
Not annealed (III)	28	253	270	0.742	38	238	238	16
1.5 hr. at 160°C	6	250	267	0.58	32			17
15 min. at 200°C	2	250	278	0.86	43			18

140

Table 6. Dynamic mechanical data of IPNs before and after annealing
sample: B-3 [60% PU (80% PCL/20% PPG)/40% PVC].

Annealing Conditions	Days After Preparing Film	Temp. at E''_{max} (K)	Temp. at $\tan \delta_{max}$ (K)	Height of $\tan \delta$ Peak	½ Width of $\tan \delta$ Peak (K)	Temp. at Another Peak or Shoulder				Figure No.
						E'' (K)	$\tan \delta$ (K)	E'' (K)	$\tan \delta$ (K)	
Not annealed	6	264	291	0.680	52					6
Not annealed	14	267	294	0.635	55					19
4 hr. at 140°C	16	273	301	0.82	49	245	245			20
8 hr. at 140°C	21	270	301	0.83	49	241	240			21
45 min. at 180°C	15	283	305	0.955	42	240	240			22
15 min. at 200°C	22	281	305	0.92	47	240	245	236	236	23
10 min. at 220°C	9	283	305	0.975	43	245	245	235	235	24
15 min. at 200°C measured again 20 days after annealing		280	313	0.62	76	243	243			25
10 min at 220°C measured again 40 days after annealing		281	313	0.605	71	244	244			26

141

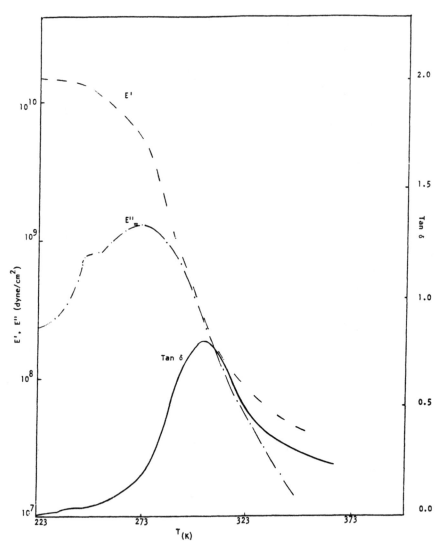

FIGURE 19. Dynamic mechanical spectrum of pseudo IPN composed of 60% polyurethane (80% PCL:20% PPG based) and 40% PVC, annealed at 140°C for 4 hours.

142

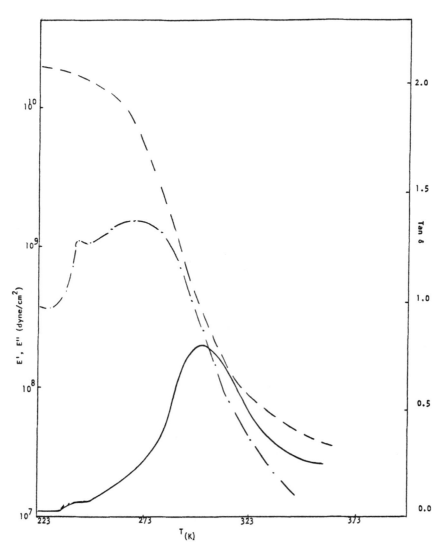

FIGURE 20. Dynamic mechanical spectrum of pseudo IPN composed of 60% polyurethane (80% PCL:20% PPG based) and 40% PVC, annealed at 140°C for 8 hours.

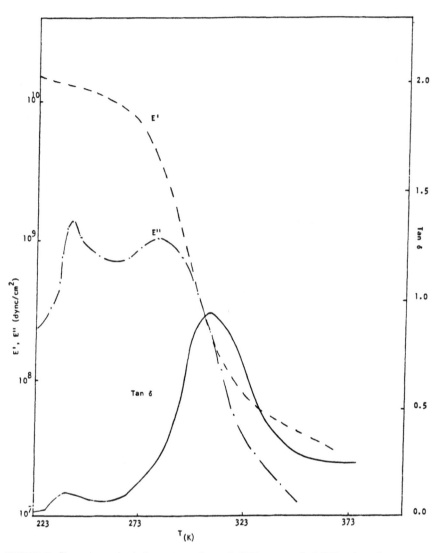

FIGURE 21. Dynamic mechanical spectrum of pseudo IPN composed of 60% polyurethane (80% PCL:20% PPG based) and 40% PVC, annealed at 180°C for 45 minutes.

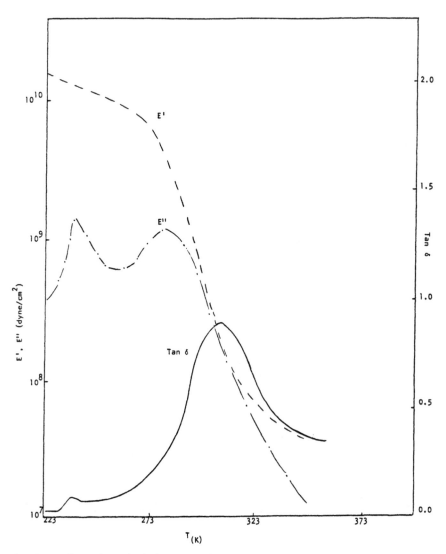

FIGURE 22. Dynamic mechanical spectrum of pseudo IPN composed of 60% polyurethane (80% PCL:20% PPG based) and 40% PVC, annealed at 200°C for 15 minutes.

145

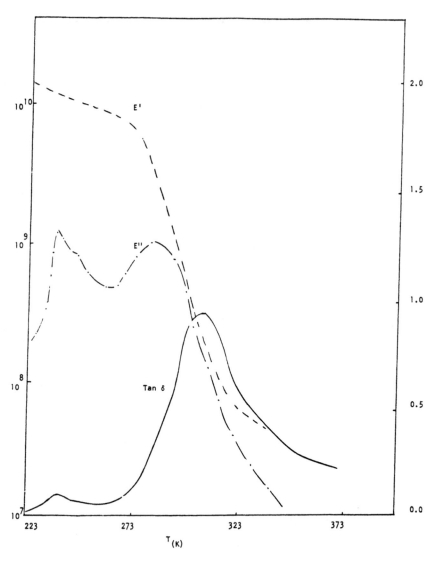

FIGURE 23. Dynamic mechanical spectrum of pseudo IPN composed of 60% polyurethane (80% PCL:20% PPG based) and 40% of PVC, annealed at 220°C for 10 minutes.

146

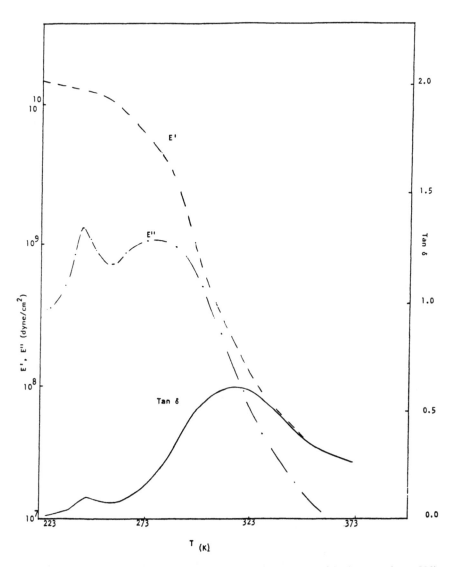

FIGURE 24. Dynamic mechanical spectrum of pseudo IPN composed of 60% polyurethane (80% PCL:20% PPG based) and 40% PVC, measured 20 days after annealing at 200°C for 15 minutes.

147

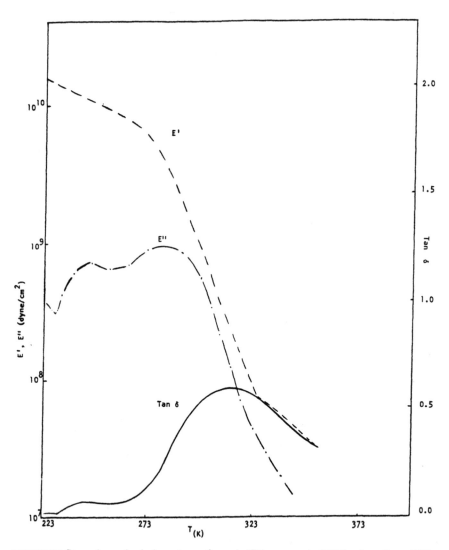

FIGURE 25. Dynamic mechanical spectrum of pseudo IPN composed of 60% polyurethane (80% PGL:20% PPG based) and 40% PVC, measured 40 days after annealing at 220°C for 10 minutes.

Table 7. Dynamic mechanical data of IPNs before and after annealing sample: B-4 [50% PU (80% PCL/20% PPG)/50% PVC].

Annealing Conditions	Days After Preparing Film	Temp. at E''_{max} (K)	Temp. at tan δ_{max} (K)	Height of tan δ Peak	½ Width of tan δ Peak (K)	Temp. at Another Peak or Shoulder E'' (K)	Temp. at Another Peak or Shoulder tan δ (K)	Figure No.
Not annealed	17	277	312	0.675	54			7
1.5 hr. at 160°C	53	290	317	0.942	45			27
45 min. at 180°C	45	294	317	0.92	47	241	243	28
15 min. at 200°C	88	295	317	0.88	47	240	242	29
5 min. at 220°C	66	293	316	0.885	37	238	243	30
15 min. at 200°C measured again 26 days after annealing		283	317	0.695	58	240	240	31

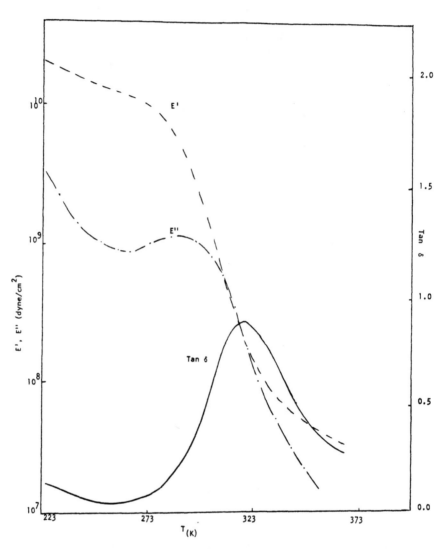

FIGURE 26. Dynamic mechanical spectrum of pseudo IPN composed of 50% polyurethane (80% PCL:20% PPG based) and 50% PVC, annealed at 160°C for 1.5 hours.

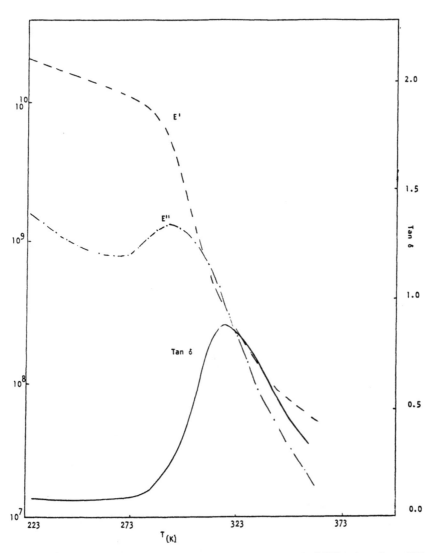

FIGURE 27. Dynamic mechanical spectrum of pseudo IPN composed of 50% polyurethane (80% PCL:20% PPG based) and 50% PVC, annealed at 180°C for 45 minutes.

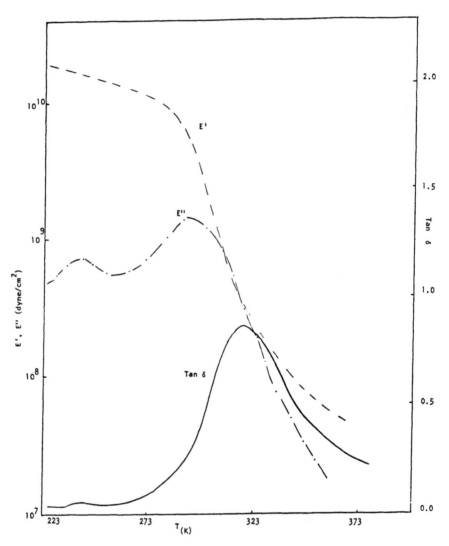

FIGURE 28. Dynamic mechanical spectrum of pseudo IPN composed of 50% polyurethane (80% PCL:20% PPG based) and 50% PVC, annealed at 200°C for 15 minutes.

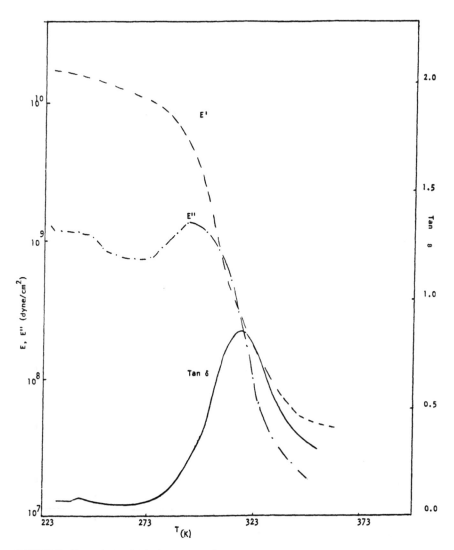

FIGURE 29. Dynamic mechanical spectrum of pseudo IPN composed of 50% polyurethane (80% PCL:20% PPG based) and 50% PVC, annealed at 220°C for 5 minutes.

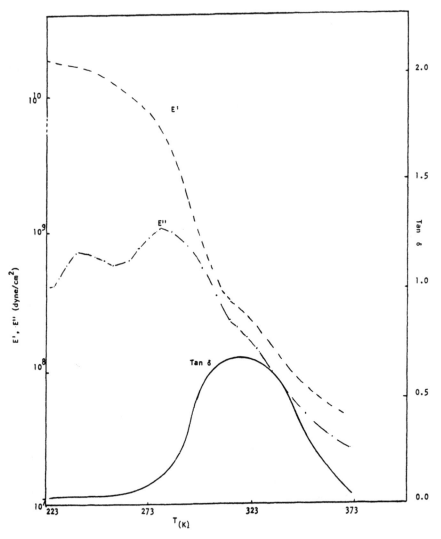

FIGURE 30. Dynamic mechanical spectrum of pseudo IPN composed of 50% polyurethane (80% PCL:20% PPG based) and 50% PVC, measured 26 days after annealing at 200°C for 15 minutes.

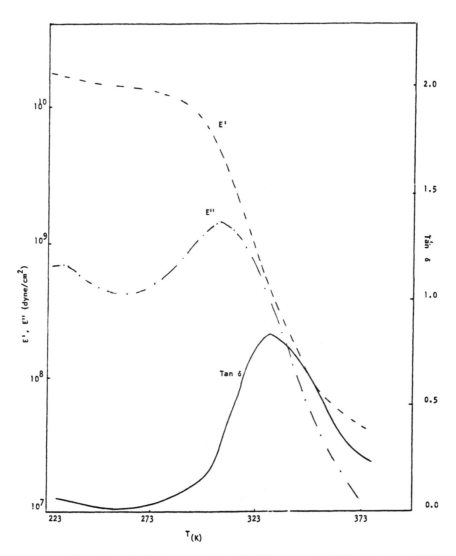

FIGURE 31. Dynamic mechanical spectrum of pseudo IPN composed of 40% polyurethane (80% PCL:20% PPG based) and 60% PVC, annealed at 200°C for 15 minutes.

155

Table 8. Dynamic mechanical data of IPNs before and after annealing
sample: B-5 [40% PU (80% PCL/20% PPG)/60% PVC].

Annealing Conditions	Days After Preparing Film	Temp. at E''_{max} (K)	Temp. at $\tan \delta_{max}$ (K)	Height of $\tan \delta$ Peak	½ Width of $\tan \delta$ Peak (K)	Temp. at Another Peak or Shoulder		Figure No.
						E'' (K)	$\tan \delta$ (K)	
Not annealed	18	286	336	0.522	86			8
15 min. at 200°C	34	305	329	0.86	51	227	225	32

156

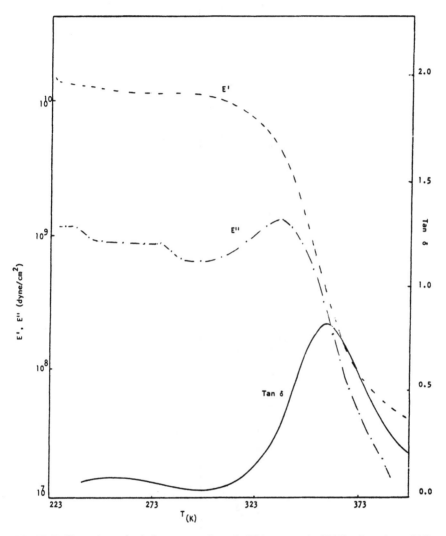

FIGURE 32. Dynamic mechanical spectrum of pseudo IPN composed of 20% polyurethane (80% PCL:20% PPG based) and 80% PVC, annealed at 200°C for 15 minutes.

Table 9. Dynamic mechanical data of IPNs before and after annealing
sample: B-6 [20% PU (80% PCL/20% PPG)/80% PVC].

Annealing Conditions	Days After Preparing Film	Temp. at E''_{max} (K)	Temp. at tan δ_{max} (K)	Height of tan δ Peak	½ Width of tan δ Peak (K)	Temp. at Another Peak or Shoulder				Figure No.
						E'' (K)	tan δ (K)	E'' (K)	tan δ (K)	
Not annealed	19	323	377	0.610	50					9
15 min. at 200°C	36	333	359	0.865	41	(279)	(279)	(232)	(235)	33

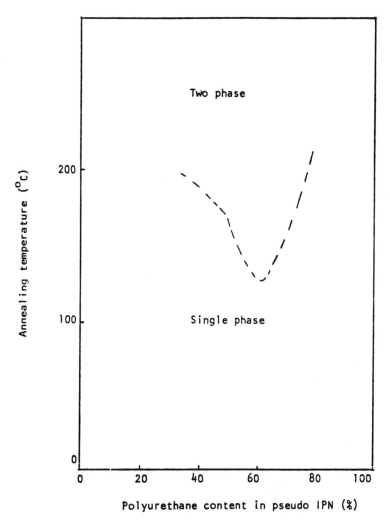

FIGURE 33. A preliminary phase diagram of pseudo IPNs composed of polyurethane (80% PCL:20% PPG based) and PVC.

The dynamic mechanical spectrum was measured again on the sample which was kept for 26 days at room temperature after annealing at 200°C (Figure 30). the main tan δ peak at 317 K (44°C) became broader as compared with that obtained soon after annealing. The data are listed in Table 7.

IPN OF POLYURETHANE/PVC: 40/60

Dynamic mechanical measurements were carried out on the sample annealed at 200°C. Figure 31 shows two tan δ (and also two E'') peaks, indicating a two-phase nature. The main tan δ peak (336 K) shifted to lower temperature (329 K) after annealing. The data are listed in Table 8.

IPN OF POLYURETHANE/PVC: 20/80

Figure 32 shows the dynamic mechanical spectrum obtained on a sample annealed at 200°C. It seems that there are a main tan δ peak at 359 K and two small peaks at 279 and 235 K. However, we cannot be sure that this spectrum indicates two-phase nature because the peaks are not clear. The data are listed in Table 9.

The data are summarized in Figure 33 (a preliminary phase diagram). Open circles show a two-phase region and filled circles show a single phase region. IPNs with high amounts of either component could not be made to phase separate at 200°C, whereas those of intermediate compositions became more heterogeneous with the thermal treatment.

ACKNOWLEDGEMENTS

The authors wish to acknowledge the contributions of Prof. Konigsveld and Dr. Kleintjens of DSM Central Lab, Geleen, The Netherlands and Prof. Harry L. Frisch of SUNY in Albany, NY for their helpful discussions. Financial support of DSM is gratefully acknowledged.

REFERENCES

1. Gesner, B. D. in *Encyclopedia of Polymer Science and Technology, Vol. 10.* H. F. Mark, N. G. Gaylord and N. Bikales (eds.). New York:Wiley-Interscience. p. 708 (1969).
2. Thompson, M. S. *Gum Plastics.* New York:Reinhold (1958).
3. Keskkula, H. *Polymer Modification of Rubbers and Plastics.* New York:Wiley-Interscience (1970).
4. Bruins, P. F. *Polyblends and Composites.* New York:Wiley-Interscience (1970).
5. Flory, P. J. *Principles of Polymer Chemistry.* Ithaca, New York:Cornell Univ. Press (1953).
6. Matsuo, M., T. K. Kwei, D. Klempner and H. L. Frisch. *Polym. Eng. Sci.,* 10:327 (1970).
7. Sperling, L. H., D. A. Thomas, II and V. Huelck. *Macromolecules,* 5:340 (1972).
8. Frisch, K. C., D. Klempner, S. Migdal, H. L. Frisch and H. Ghiradella. *Polym. Eng. Sci.,* 14:76 (1974).
9. Frisch, L. H. and E. Wasserman. *J. Am. Chem. Soc.,* 83:3789 (1961).
10. Frisch, K. C. and D. Klempner. *Adv. Macromol. Chem.,* 2:149 (1970).
11. Flory, P. J. *Chem. Rev.,* 35:51 (1944).

12. Prager, S. and H. L. Frisch. *Chem. Phys.*, 46:1475 (1967).

13. Curtius, A. J., M. J. Covitch, D. A. Thomas and L. H. Sperling. *Polym. Eng. Sci.*, 12:101 (1972).

14. Klempner, D., H. L. Frisch and K. C. Frisch. *J. Polym. Sci. A-2*, 8:921 (1970).

15. Klempner, D., H. L. Frisch and K. C. Frisch. *J. Elastoplastics*, 3:2 (1971).

16. Frisch, K. C., D. Klempner, S. Migdal and H. L. Frisch. *J. Polym. Sci. A-1*, 12:885 (1975).

17. Frisch, K. C., D. Klempner, S. Migdal and H. L. Frisch. *J. Appl. Polym. Sci.*, 19:1893 (1975).

18. Frisch, K. C., D. Klempner, S. K. Mukherjee and H. L. Frisch. *J. Appl. Polym. Sci.*, 18:689 (1974).

19. Frisch, K. C., D. Klempner, T. Antczak and H. L. Frisch. *J. Appl. Polym. Sci.*, 18:683 (1974).

20. Kim, S. C., D. Klempner, K. C. Frisch, H. L. Frisch and H. Ghiradella. *Polym. Eng. Sci.*, 15:339 (1975).

21. Klempner, D. and K. C. Frisch. *Adv. Urethan. Sci. Technol.*, 3:14 (1974).

22. Kim, S. C., D. Klempner, K. C. Frisch and H. L. Frisch. *J. Appl. Polym. Sci.*, 21:1289 (1977).

22a. Klempner, D. Unpublished.

23. Frisch, K. C., D. Klempner, H. L. Frisch and H. Ghiradella in "Review Article," *Recent Advances in Polymer Blends, Grafts, and Blocks*, L. H. Sperling (ed). New York:Plenum Press (1974).

24. Sperling, L. H., D. A. Thomas, H. E. Lorenz and E. H. Nagel. *J. Appl. Polym. Sci.*, 19:2225 (1975); J. A. Graves, D. A. Thomas, E. C. Hickey and L. H. Sperling. *J. Appl. Polym. Sci.*, 19:1731 (1975).

25. Frisch, K. C., H. L. Frisch, D. Klempner and S. K. Mukherjee. *J. Appl. Polym. Sci.*, 18:689 (1964); K. C. Frisch, D. Klempner, T. Antzak and H. L. Frisch. *ibid.*, 18:683 (1974); K. C. Frisch, K. Klempner, S. Migdal, H. L. Frisch and H. Ghiradella. *Polym. Eng. Sci.*, 15:339 (1975).

26. Sperling, L. H. and R. R. Arnts. *J. Appl. Polym. Sci.*, 15:2731 (1971); R. E. Touhsaent, D. A. Thomas and L. H. Sperling. *J. Polym. Sci.*, 46C:175 (1974).

27. Sperling, L. H., D. A. Thomas and V. Huelck. *Macromolecules*, 5:340 (1972).

28. Klempner, D., C. L. Wang, M. Ashtiani and K. C. Frisch. *J. Appl. Polym. Sci.*, 32:4197 (1986).

29. Unger, E. E. in *Noise and Vibration Control*. L. L. Beranek (ed.). New York:McGraw-Hill (1971).

30. Kalfoglou, N. K. *J. Appl. Polym. Sci.*, 26(3):823 (1981).

31. Hoursten, D. J. and I. D. Hughes. *J. Appl. Polym. Sci.*, 26(10):3467 (1981).

32. Garcia, D. *J. Appl. Polym. Sci., Part B: Polym. Physics*, 24:1577 (1986).

33. Ball, G. L. and I. O. Salyer. *J. Acoust. Soc. Amer.*, 39:663 (1966).

34. Wang, C. B. and S. L. Cooper. *J. Appl. Polym. Sci.*, 26:2982 (1981).

35. Balga, A. and D. Feldman. *J. Appl. Polym. Sci.*, 28:1033 (1983).

36. Piglowski, J., T. Skowronski and B. Masiulanis. *Angew. Makromol. Chem.*, 85:129 (1980).

37. Shen, C. H. and Y. Y. Wang. *Angew. Makromol. Chem.*, 21:49 (1984).

38. Piglowski, J. and W. Laskawski. *Angew. Makromol. Chem.*, 82:157 (1979).

39. Klempner, D., K. Hibino and K. C. Frisch. *J. Polymer Sci.* (in press).

11

Engineering Rheology in the Design and Fabrication of Industrial Polyblends

A. P. PLOCHOCKI*

THE ENGINEERING RHEOLOGY OF THE
POLYBLENDS/ALLOYS: GENERAL [4,10b,21]

THE UNDERPINNING ASSUMPTION adopted here is that the material functions (shear viscosity, elasticity, elongation viscosity and the derived characteristics) which originate from constitutive equations for the continuum are applicable to the multiphase melt flow description. The convenience of using the same description procedures for the component—polymers and for the polyblends— has to be compensated, therefore, by a number of strong simplifying assumptions and semiempirical correlations. Most of the simplifications follow from neglecting coupled effects of normal stresses and the interface tension on the concomitant phase structure (morphology) of a polyblend. Perhaps an equally crude approximation is in neglecting the effect of isotropic pressure: there is some evidence that the pressure dependence of the melt rheology of the polyblends [55] is even stronger than that observed for polymers [5–7].

With an understanding of the above limitations, it is possible to develop a coherent, engineering approach in selecting components and composition of the polyblends, to estimate their morphology and set up the mixing ("compounding") process as well as to assess the effects of processing on the morphology and the performance characteristics. Basic characteristics of melt rheology are grouped in Table 1; in view of the assumptions made it does not appear justifiable to use a more elaborate approximation for the material functions than those of the power law [see Equations (1) and (2) in Table 1] and the log parabola [see Equations (5–7) in Table 1]. It has to be underscored, however, that the functions' approximations are valid within the stress range specified in the database [2,3a]. It is also essential to realize that the melt rheology of the polyblends/alloys is time

* 2305 Mountain Ave., Scotch Plains, NJ 07076.

Table 1. Melt rheology—the simplest approximations for the material functions.

Viscosity	Elasticity — General	Elasticity — High Shear Estimates
shear stress	elasticity coeff.:	
	principal normal stress	
(1) $p_{12} = p_{12}^0 \cdot \left(\dfrac{\dot\gamma}{\dot\gamma^0}\right)^n$	(2) $N_1 = N_1^0 \cdot \left(\dfrac{\dot\gamma}{\dot\gamma^0}\right)^m$	$\beta = \left[1 + 0.5\left(\dfrac{N_1}{p_{12}}\right)^2\right]^{1/6}$
	$\psi_1 = \dfrac{N_1}{\dot\gamma^2}$	(Tanner)
viscosity function: power law—	elasticity function:	recoverable strain:
	relaxation time:	
(3) $\eta = \eta_0 \cdot \dot\gamma^{n-1}$	(4) $\psi_1 = \psi_1^0 \cdot \gamma^{m-2}$	$\gamma = \dfrac{N_1}{2p_{12}} = \sqrt{2(\beta^6 - 1)}$
	$\bar\tau = \dfrac{\psi_1}{2\eta}$	
or log parabola approximations	relaxation time function	$\gamma = $ const. p_{12}^b (for b < 1, (l/d) < 40)
(5) $\ln\eta = A0 + A1\ln\dot\gamma + A2\ln^2\dot\gamma$	(6) $\ln\psi_1 = B0 + B1 \cdot \ln\dot\gamma$	$\gamma = \sqrt{(b^2 + 1)(\beta^6 - 1)}$
	(7) $\ln\bar\tau + C0 + C1 \cdot \ln\dot\gamma + C2\ln^2\dot\gamma$	
standardized viscosity:	standardized elasticity:	(strain on the capillary wall)
$\eta^0 = |p_{12}^0|$ or e^{A0}	$\psi_1^0 = e^{B0}$	$F = \gamma_w = \dfrac{3n + 1}{2(5n + 1)}$
	standardized relaxation time	
	$\bar\tau^0 = e^{C0}$	
power law (viscosity) exponent	elasticity exponent:	(Utracki)
$n \sim A1 + 1$	$m = B1 + 2$	

At standardized shear rate $\dot\gamma^0$ of 1 sec⁻¹.

dependent: the domain-shape recovery and coalescence driven processes (Figure 1) tend to change the morphology of the liquid polyblend and thereby its rheological response. The kinetics of the change are temperature and stress field dependent hence the time scale and the stress level of the industrial processes has to be comeasurable with these used in melt rheometry. In particular caution should be exercised in using dynamic rheometry data for the design purposes: the Merz-Colwell rule rarely (if at all) holds, and vibration enhances the coalescence.

Although both physical compatibilization (incorporation of a block copolymer [49,59]) and chemical compatibilization (by a chemical reaction on the interface) slow down the changes in morphology of a polyalloy (= polyblend + compatibilizer), the rheological response of a polyalloy to the time scale of a process will, as a rule, be stronger than that of a polymer melt.

ENGINEERING MELT RHEOLOGY OF INDUSTRIAL POLYBLENDS: FACTORS CONTROLLING TYPE AND THE DEGREE OF DISPERSION OF THE MINOR PHASE (MORPHOLOGY), SYNERGISM OF THE MELT VISCOELASTICITY AND EVALUATION OF THE SELECTED "MIXING RULES"[1]

Overview

In the binary mixtures of commercial polymers ("polyblends") the composition dependence of melt rheology is inherently nonlinear. A synergism in the (shear) viscosity and elasticity as well as in elongational viscosity is typically observed at composition(s), "particular" for the polyblend under consideration [1a,3]. Standardized treatment of melt rheology data [2] for a series of blends consisting of the same components and differing in composition yields series of viscosity, elasticity, etc., curves which differ in slope and display crossovers [3] (Figure 2). From a formal point of view, therefore, the "particular rheology" composition simply reflects curve crossovers at a selected shear level (Figure 3). It is interesting to note that the "parameters" of the normalized presentation [2,17] of polyblends melt rheology viz. standardized viscosity (η^0); elasticity (ψ_1^0); and the respective (Table 1) power laws' exponents, n and m, plotted against composition form mirror-image (property-composition—Figure 4) curves.

Since the slope of a material function curve is strongly dependent on supermolecular structure even in the melts of neat polymers [4], the association of PRC with changes in the phase structure (morphology) seems to be justified. In the early 1970s, Simha (see [3]) suggested a correspondence between the specific melt volume, v_T and melt viscosity of a polyblend. Simple measurements of v_T during the capillary flow (Schreiber, Rudin, 1970) corrected for the differences in extrusion pressures (Δp) at the particular flow rate levels [by considering

[1]Presented in part at the Annual Meeting of the Society of Rheology, Knoxville, TN, June 1983 (paper E6).

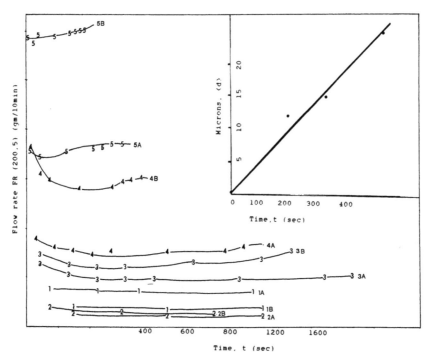

FIGURE 1. Time dependence of melt fluidity (FR 200.5A-ASTM D-1238) for the polyblends (B) and polyalloys (A): 1,2 ≅ 67; 3 ~ 85 and 4,5–33 (expressed in wt. % of LDPE) ("composition"). Upper right corner: same dependence of polystyrene domain size in LDPE/PS = 2/1 polyblend at 473 K (courtesy Diane M. Kidwell and John E. Curry).

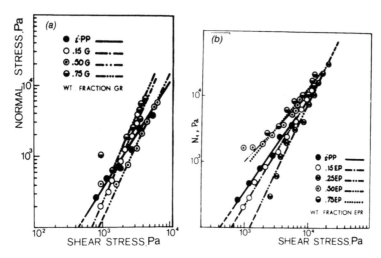

FIGURE 2. The elasticity functions $N_1(p_{12})$ of the rubbers' polyblends and their components—the composition dependence and crossovers: (a) i-PP/natural (GR) rubber; (b) i-PP/EP rubber [78].

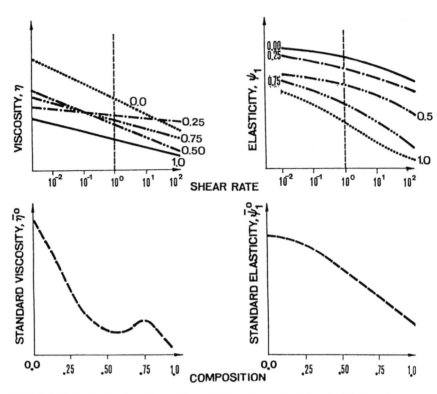

FIGURE 3. Particular Rheology Compositions (PRC) as the result of viscoelasticity functions cross-over: standardized shear rate, $\dot{\gamma}^0$ is the reference abscissa.

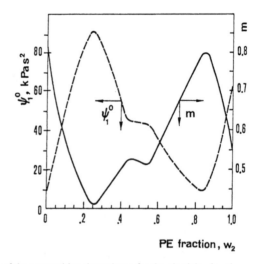

FIGURE 4. Pattern of the composition dependence for the elasticity function parameters, ψ_1^0 and m.

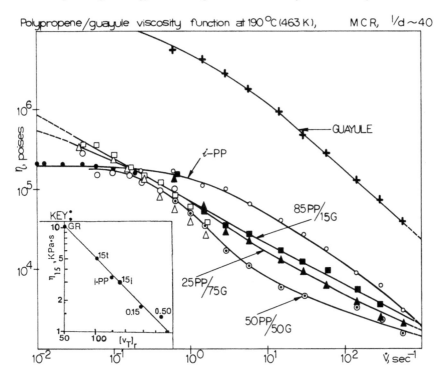

FIGURE 5. Melt viscosity, η_{15}—specific volume, (v_T), dependence for i-PP/GR polyblends, the data for different compositions and distributive mixing processes. Viscosity curves (463 K): 15*i,t*—the first and last specimens, static mixer; 15,50 "Banbury" prepared specimens of the polyblend containing 15 and 50 wt.% of GR [78].

$(v_T)_r = v_T/\Delta p$, i.e., the melt compressibility rather than v_T] indicate a close correspondence[2] with shear viscosity, η (Figure 5).

Evidently, melt fluidity synergism reflects less dense packing of phases or perhaps even a transition from disperse into interpenetrating and/or stratified arrangement of phases ("domains") of components. The hypothesis of the packing-density-change, however, provides only a partial explanation for the transition in the morphology at PRC. Composition at which the synergistic melt rheology appears to be the same for both viscosity *and* elasticity functions in spite of the fact that the elasticity was measured at the simple shear, and at a much lower pressure. The complexity of the melt rheology at PRC is further augmented by the difference between the polyblend morphology concomitant to

---

[2]For a comprehensive treatment of the pressure dependence of polymers viscosity, see Utracki [5,41], Litt [6] and Semjonov [7]. Linear relationship between v_T and shear viscosity, η, observed here invalidates the "criterion" of applicability for the continuum hypothesis for molten polyblends. The criterion based on the overlapping of the cone-plate and the capillary rheometry data requires the overlap to occur for both viscosity and elasticity functions ($N_1 - N_{1B}$ form [47]).

capillary flow (bimodal disperse) and that concomitant to the simple shear (typical monodisperse—Figures 6 and 7). It appears, therefore, that at PRC more than one type of melt morphology is present. However, Porter [8], White [9], Han [10], Doppert [11] and others [12] indicate that the core/sheath morphology is fairly common. Current theories of phase stability, e.g., [1b,13,21,29b] are concerned with changes in distribution of phases in the quiescent melt, interpreted in terms of spinodal decomposition [13].

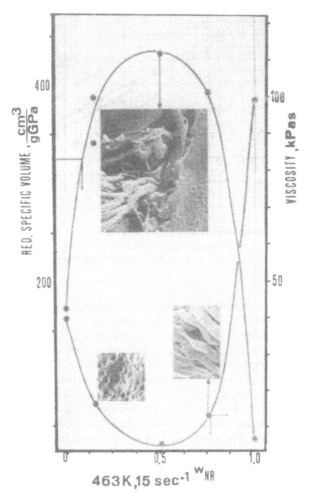

FIGURE 6. Morphology at the "synergistic" composition (PRC), i-PP/GR polyblends: coarse core/ sheath system at 50 wt.%. The viscosity and specific volume data points are connected with the trend lines.

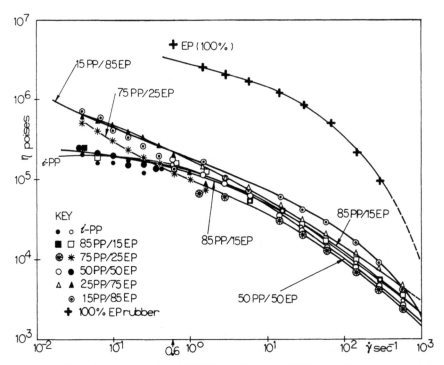

FIGURE 7. Viscosity functions of the i-PP/EPR polyblend: the 75/25 composition is "synergistic."

Descriptions of the morphology of *sheared,* molten polyblends, attempted by VanOene [15] and Jorgensen [14] are closer to the morphological mechanism of PRC, albeit composition effects are still to be accounted for. In particular the VanOene disperse-stratified transition in morphology could well be "responsible" for formation of the core-sheath phase distribution observed at PRC.

Intensity of a particular composition [e.g., amplitude of the standardized viscoelasticity functions: $(\eta^0)_{max}$ or $(\psi_1^0)_{max}$] and its placement also depend on two other factors:

- The polyblend fabrication, e.g., the distributive and/or dispersive melt mixing process. Assuming that the VanOene transition is the underlying mechanism for PRC, we have yet to account for the fact that the transition takes place only at the domain size exceeding $<d> = 1$ μm
- The polyblend processing which may be represented (at a temperature) by stress field intensity. This is illustrated with shear dependence of the synergistic "elasticity" composition for a mixture of polyethylenes (*d* in [3a]). With increasing shear the composition at which $(\psi_1)_{max}$ (i.e., PRC—Figure 8) is observed, is also increasing. In the same time there is a decrease in the #1 PRC "intensity," i.e., in the height of $(\psi_1)_{max}$ peak. The trend continues until after a certain shear level is attained: beyond this level PRC (#1) shifts to a

lower concentration (PRC #2 appears) and, with increasing shear we again observe the decreasing PRC (#2) intensity.

The Viscosity and Elasticity Ratio Functions of the Binary Polyblends—Dependence on Shear Stress (p_{12} [3c])

In a polyblend consisting of i and the m ("inclusion" and "matrix") components of viscosities η_i, η_m and elasticities $_i\psi_1$, $_m\psi_1$ the effectiveness of dispersive (shear) mixing depends on the difference in the "inherent" shear rates, $\Delta\dot{V} = \dot{V}_i - \dot{V}_m$ which is instrumental in the "rheological" compatibilization [16,20,45]. Too large a shear difference $\Delta\dot{V}$ precludes an interaction of the components and results in a weak interface. The melt viscoelasticity ratio functions [50]: η_r, $(\psi_1)_r$ at a specific level of $\Delta\dot{V}$, constitute, therefore, an important criterion, the selection of component polymers. Shear dependence of melt viscosity ratio, η_r is illustrated with Figure 9. Assuming power law dependence of η_r and $(\psi_1)_r$ on $\dot{V}$ with the obvious [15,16] stipulation that shear stress is the independent variable ($p_{12} = _ip_{12} = _mp_{12}$) we have:

$$p_{12} = p_{12}^0 \, (\dot{V}/\dot{V}^0)^n \tag{1a}$$

with

$$\dot{V}^0 = 1 \text{ sec}^{-1} \text{ and } \psi_1 = \psi_1^0 \, (\dot{V}/\dot{V}^0)^{m-2} \tag{1b}$$

where

$$N_1 = N_1^0 \cdot (\dot{V}/\dot{V}^0)^m \tag{1c}$$

and

$$\psi_1 = N_1 \cdot \dot{V}^{-2} \tag{1d}$$

are first normal stress difference and elasticity, respectively. Considering *integer* equalities:

$$|\psi_1^0| = |N_1^0| \text{ and } |p_{12}^0| = |\eta^0| \text{ with } \eta = \eta^0 \cdot \dot{V}^{n-1} \tag{2}$$

and defining the viscoelasticity ratio functions:

$$VR = \eta_r = \eta_i/\eta_m \text{ and } \eta_r^0 = \frac{\eta_i^0}{\eta_m^0} \tag{3a}$$

$$ER = (\psi_1)_r = \frac{_i\psi_1}{_m\psi_1} \text{ and } (\psi_1^0) = {_i\psi_1^0}/{_m\psi_1^0} \tag{3b}$$

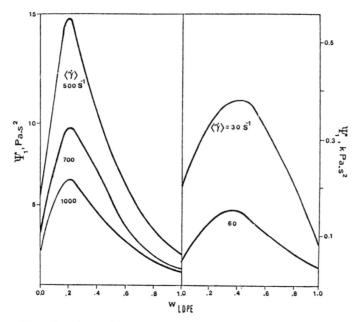

FIGURE 8. Shear dependence of intensity and on the composition axis of the location elasticity synergism for HDPE/LDPE polyblends. (Apparent flow rates are annotated on the curves.)

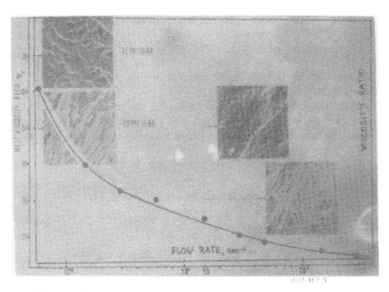

FIGURE 9. Shear rate dependence of the viscosity ratio, η_r and concomitant morphologies for the i-PP/GR polyblends—the experimental data. Note VanOene transition in the type of morphology (e.g., platelets of i-PP in the rubber matrix).

We then arrive at the shear stress dependent viscosity ratio function:

$$VR = \eta_r = \eta_r^0 \cdot \frac{[(p_{12}/\eta_m^0)_m^{1/nm} + \Delta\dot{V}]^{n_i-1}}{\left(\dfrac{p_{12}^0}{\eta_m^0}\right)^{(nm-1)/nm}} \quad \text{for } \Delta\dot{V} = \dot{V}_i - \dot{V}_m \quad (4a)$$

$$VR' = \eta_r' = \eta_r^0 \cdot \frac{\left(\dfrac{p_{12}}{\eta_m^0}\right)^{(n_i-1)/n_i}}{\left[\left(\dfrac{p_{12}}{\eta^0}\right)^{1/n_i} - \Delta'\dot{V}\right]^{(nm-1)}} \quad \text{for } \Delta'\dot{V} = V_m - \dot{V}_i \quad (4b)$$

Alternatively (Mekkaoui [35]) one may introduce $q = 1/n$ for the reciprocal of the power law exponent, and $S = (p_{12}/\eta^0)$ for ratio of shear stress and standardized melt viscosity, $k = (m - 2)/n$ for the (adjusted) ratio of elasticity and viscosity power law exponents (*m* and *n*).

The viscosity ratio function then becomes:

$$VR = \eta_r = \eta_r^0 [S_m^{qm} + \Delta\dot{V}]^{(n_i-1)} \cdot (S_m)^{(1-nm)/nm} \quad (5a)$$

or

$$VR = \eta_r = \eta_r^0 [S_i^{qi} + \Delta\dot{V}]^{(1-nm)} \cdot (S_i)^{(n_i-1)/n_i} \quad (5b)$$

which correspond to (4a) and (4b). In a simplified form the function is:

$$VR = \eta_r = S_i^{-1} \cdot (S_i^{qi} - \Delta\dot{V})^{nm} \quad (5c)$$

In Figures 10–12 the viscosity ratio function is presented for selected polyolefins blends.

Melt elasticity of the polyblend, in particular, the principal difference in normal stresses ($\Delta N_1 = {}_mN_1 - {}_iN_1$) and the interface tension $\nu_{i,m}$ constitute a feasibility criterion for the transition between disperse and stratified morphologies [15].

The stress dependent elasticity is:

$$N_1 = N_1^0 \cdot S^* = \psi_1^0 \cdot (\eta^0)^{-k} \cdot p_{12}^k \quad (6)$$

and "melt elasticity ratio" (MER) is:

$$(\psi_1)_r = (\psi_1^0)_r \cdot (\eta_m^0)^{km} \cdot (\eta_i^0)^{-ki} \cdot (p_{12})^{\Delta k} \quad (7)$$

where $\Delta k = k_i - k_m$. Δk may be used in predicting shear induced changes in phase distribution (disperse or stratified arrangement). In an analogy to the engineering applications of the viscosity ratio function, $\eta_r(p_{12})$ we have to

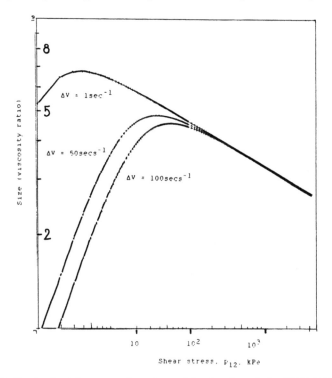

FIGURE 10. The simulated shear stress dependence of the viscosity ratio for HDPE/LDPE (type "d" [3a]) polyblends; ΔV—the arbitrary differences in inherent shear rates (courtesy B. H. Luor).

account for the difference in the inherent shear rates within the i and m component-polymers phases. Designating the difference by $\Delta \dot{V}$ we have:

$$ER = (\psi_1)_r = (\psi_1^0)_r \cdot (S_m)^{-km} \cdot [S_m^{qm} + \Delta \dot{V}^{(m_i-2)}] \tag{8a}$$

for $\Delta \dot{V} = \dot{V}_i - \dot{V}_m$, and:

$$(\psi_1)_r = (\psi_1^0)_r \cdot (S_i)^{ki} \cdot (S_i^{qi} - \Delta' \dot{V})^{(2-m_m)} \tag{8b}$$

for $\Delta' \dot{V}_m = \dot{V}_m - \dot{V}_i$. Since it is convenient to use the ratio rather than the difference in inherent shear rates, the ratio $g = \dot{V}_i / \dot{V}_m$ is usually substituted for $\Delta \dot{V}$ in specification of the commercial polyblends *and* in the description of the melt mixing processes. The elasticity ratio, MER, therefore, would have the form:

$$(\psi_1)_r = (\psi_1^0)_r \cdot (g)^{ki} \cdot (S_m)^{(gm \cdot ki-km)} = (\psi_1^0)_r \cdot g^{ki} \cdot (\psi_m^0)^{(km-qm \cdot ki)} \cdot p_{12}^{(qm \cdot ki-km)} \tag{9}$$

Since the first three terms of Equation (9) are for a selected polyblend and a given mixing process, a constant ($= \chi$), putting ($q_m \cdot k_i - k_m) = e$ we may have Equation (9) in a form convenient for evaluating shear effects in the system under consideration:

$$ER = (\psi_1)_r = \chi \cdot p_{12}^e \tag{10}$$

Considering typical values of m, n exponents (see Tables 1–3 in [3a]) it can be shown that $e \cong 2$ and, therefore, that MER is a rather strong function of shear stress. Shear dependence of (melt) elasticity ratio is illustrated with Figures 13–15 for three sets of components of the polyblends. This observation confirms that the phase distribution (morphology) of a molten polyblend is a resultant of the shear stress field. To specify the melt rheology of components polymers in terms of their Newtonian viscosities, η_0 (as, e.g., in [18]) leads, therefore to erroneous predictions for the type of morphology of the immiscible polyblends, concomitant to the mixing *and* processing operations.

Typical traces of the shear stress dependent viscosity ratio, *VR* and elasticity ratio, *ER* for i-PP/HDPE polyblend at 463 K [3,19] are given in Figures 12 and 10. While changes in the viscosity ratio function are moderate with a maximum

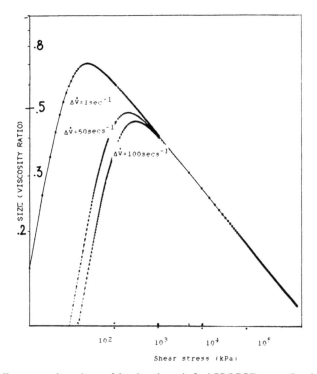

FIGURE 11. Shear stress dependence of the viscosity ratio for i-PP/LDPE (type "a") polyblends ([3a] and Dr. S. K. Dey).

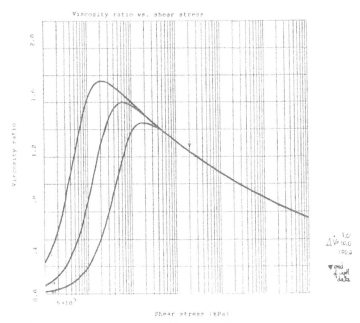

FIGURE 12. Shear stress dependence of viscosity-ratio for i-PP/HDPE (AsHos) polyblends ([3a] and B. H. Luor [35]).

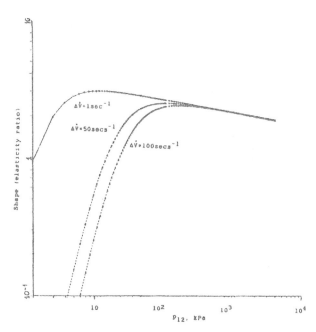

FIGURE 13a. Shear stress dependence of the elasticity ratio (ψ_1), for the "d" HDPE/LDPE polyblends ([3a] and B. H. Luor [35]).

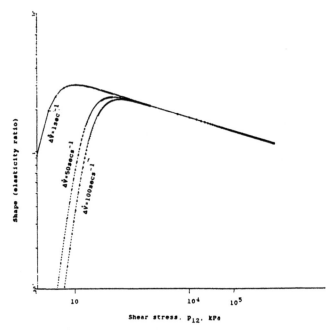

FIGURE 13b. Ditto for the upper confidence limit set of data [3a].

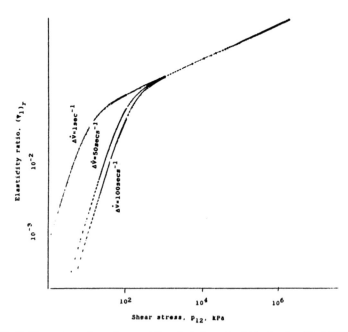

FIGURE 14. Shear stress dependence of the elasticity ratio (ψ_1)$_r$ for the "a" (i-PP/LDPE [3a] polyblends) [35].

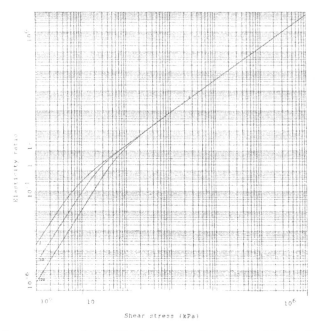

FIGURE 15. Shear stress dependence of the elasticity ratio for "AsHos" polyblends ([3a] and B. H. Luor [35]).

at the stress level close to $|\eta^0| = |p_{12}^0|$, changes in the elasticity ratio *ER* extend over 8 orders of magnitude (10^{-5} to 10^3). The experimental range of shear stress extends from 1 to 1250 kPa: its upper limit is indicated with ▲. To verify whether the observed $VER - p_{12}$ dependencies form a pattern, data for HDPE/LDPE (*d*) and i-PP/LDPE (*a*) [3a] systems were presented in Figures 10, 13, 11 and 14, respectively. While the stress dependence of the viscosity ratio, η_r indeed follows the pattern (ranges of η_r values are 1 to 6.5 for *d*, and 0.15 to 0.70 for *a* polyblends), the elasticity ratios $(\psi_1)_r$ are considerably less dependent on shear stress, p_{12} [the ranges of $(\psi_1)_r$ are 0.9 to 3.0 for *d* and 10^{-3} to 2.0 for *a* polyblends]. In both cases the stress dependence is significant. Within the lower end of the shear stress range it clearly depends on the difference in "inherent" shear rates $\Delta \dot{V}$. These were arbitrarily set at 1, 10 and 100 sec^{-1}. The relationships of η_r and $(\psi_1)_r$ with size and shape, respectively, of the *minor* phase domains have yet to be quantitatively developed. Development might be feasible by accounting for the composition effect: the interchange of the inclusion-matrix roles between the components in existing theories [14,43,51]. In other words, both η_r, $(\psi_1)_r = f(p_{12})$ and

$$\frac{1}{\eta_r}, \frac{1}{(\psi_1)_r} = f_1(p_{12})$$

have to be considered in parallel to the effects of a qualitative change in the morphology associated with PRC. With respect to the latter requirement, VanOene's [15] and Dey/Bendaikha's [27] approach appear promising. As an example of the current approach to the problem the procedure of Martinez and Williams [20] is fairly typical. They estimate the reduction of the "domain" (minor phase particles) size $<d_i>$ in shear on the basis of the Taylor-Oldroyd theory developed for dilute emulsions with the Newtonian matrix. Assuming that this approach is valid in the case of molten polyblends in shear, one arrives at an expression for the "overall" viscosity:

$$\eta = \eta_m \left(1 + \frac{1 + 2.5\eta_r}{1 + \eta_r} \right) \cdot C_i \tag{11}$$

where

η_m = viscosity of the major (matrix) phase
η_r = viscosity ratio
C_i = minor phase (inclusion) mass fraction

The principal normal stresses difference is:

$$N_1 = \frac{\nu_{i,m}^{-1}}{40} \cdot \eta_i^2 \cdot <d_i> \cdot \left(\frac{19\eta_r + 16}{\eta_r + 1} \right)^2 \cdot \left(\frac{p_{12}}{\eta^0} \right)^{2q} \tag{12}$$

where $\nu_{i,m}$, the interface tension is given by:

$$\nu_{i,m} = \nu_i + \nu_m - 2\,C_i \cdot \nu_i \cdot \nu_m \tag{12a}$$

Estimates for η and N_1 could obviously be combined to give an effective relaxation time:

$$<\tau> = \frac{\eta_i \cdot <d> \cdot C_i}{\nu_{i,m}} \cdot f(\eta_r) \cdot \frac{N_1 \cdot (\eta^0)^{-q}}{2\eta_i \cdot p_{12}^q} \tag{13}$$

The ratio of (standardized) relaxation times τ_r^0 for a molten polyblend represents "shear toughness" of the minor phase (see Figure 16 and [3a]). At the same time, the "relaxation time ratio," τ_r^0 embodies merits of effective relaxation time at the specified shear level. Therefore, it has an advantage over Williams' "effective relaxation time," τ_{eff} in specifying the mean domain size, $<d_i>$ by:

$$<d_i> = \frac{(\tau^0)_r \cdot \nu_{i,m}}{C_i \cdot \eta_i \cdot f(\eta_r)} \tag{14}$$

An alternate estimate for $<d_i>$, derived independently by Martinez/Williams [20] and by Han [21] contains the explicit (Taylor-Oldroyd) form of the viscosity

Characteristic Relaxation Time as a Measure of Molten Blends' Deformability

An alternative way for evaluating composition dependence of blends viscoelasticity using standardized viscosity, η^0, elasticity, ψ_1^0, and characteristic relaxation time, $\bar{\tau}^0$, where

$$\bar{\tau}^0 = \frac{\psi_1^0}{2\eta^0} \tag{1}$$

and

$$\psi_1^0 = N_1^0 / \dot{\gamma}^2, \quad p_{12}\text{-shear stress and } \dot{\gamma}\text{-shear rate} \tag{2}$$

is to use reduced rheology characteristics (e.g., $\bar{\tau}^0$), expressed as the ratio of a characteristic for a blend, $\bar{\tau}_b^0$, divided by the characteristic of a component, $\bar{\tau}_1^0$.

Assuming validity of the relationship

$$N_1 = 2p_{12}\gamma, \quad \text{where } \gamma\text{-recoverable strain} \tag{3}$$

At the standard shear rate, $\dot{\gamma}^0 \equiv 1 \text{ s}^{-1}$, one has:

$$\bar{\tau}_r^0 = \frac{\bar{\tau}_b}{\tau_1^0} = \frac{(N_1^0)_b/(N_1^0)_1}{(\eta^0)_b/(\eta^0)_1}$$

$$= \frac{(p_{12}^0 \cdot \gamma^0)_b}{(p_{12}^0 \cdot \gamma^0)_1} \cdot \frac{(p_{12}^0)_1}{(p_{12}^0)_b} = \frac{\gamma_b^0}{\gamma_1^0} = \gamma_r^0 \tag{4}$$

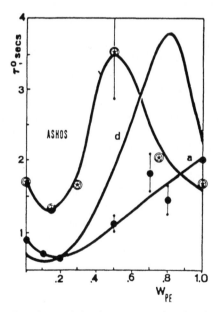

FIGURE 16. Composition dependence of the characteristic relaxation time for the three types of polyblends [3a] and a deviation of the reduced relaxation time, τ_r^0.

ratio function $f(\eta_r)$ and it may be written in terms of parameters defined previously, as:

$$<d_i> \; = \; \frac{40 \, N_1 \cdot v_{i,m} \cdot (\eta_r + 1)^2 \cdot (\eta^0)^{-2q}}{(19\eta_r + 16)^2 \cdot \eta_m \cdot p_{i\tilde{2}}^{2q} \cdot v_i} \tag{15}$$

with v_i – vol. fraction of the "inclusion" (minor component). In Equation (15) symbols without subscripts refer to the polyblend characteristics specified at a shear stress level.

The relationships between $<d_i>$ and the *VER* functions were successfully verified experimentally for the "model" PE/PS polyblend and a number of industrial mixing processes [80]. Specifically, the effects of the functions on the shape of the minor phase and on the VanOene transition at PRC are of interest. Available data however already confirm applicability of Equation (15) to polyblends ranging from high impact rubber containing polyblends to these with liquid crystal polymers.

Predicting Properties of the Polyblends Beyond the Range of Synergistic Compositions: The Example of "Mixing Rules" for the Shear Viscosity of Binary Mixtures[3]

Attempts to formulate reliable predictions for (shear) viscosity of mixtures which display nonlinear dependence of properties on composition were made in the late 19th century [22]. Empirically selected means which would fit experimental viscosity-composition relationships for binary mixtures of weakly interacting liquids were used.

The simplicity of such an approach appeals even today to rheologists who use and/or rederive (not always correctly) these rules for the polyblends-viscoelastic systems of prominent structural, rheological and interaction complexity. Development of adequate mixing rules for polymer/polymer mixtures was pioneered by Bogue [23,24] and Carley [25] who introduced 2nd and 3rd order "interaction" terms into the classic mixing laws. In a simplified manner, therefore, the effects of the mixing process were taken into account. On the other hand, the effect of processing flow was incorporated into a mixing rule by Kasajima et al. [26] in the form of a coupled composition-shear interaction factor, or the "mixing effectiveness indicator" (MEI).

The significance of the "interaction" and "shear effects" terms for applications of the mixing rule in polyblends' engineering becomes evident after reviewing the principal rules developed for ordinary mixtures.

For a non-additive-viscosity mixture of Newtonian liquids Arrhenius proposed the viscosity-composition relationship:

$$\eta = \eta_i^x \cdot \eta_m^{(1-x)} \tag{16}$$

---

[3]Newer approach of the "experiments" [48] extends the mixing rule concept to three and more component polyblends.

where x is mass fraction and $\eta_{i,m}$ are viscosities of the components. Arrhenius realized the limitations of Equation (16) stating that it should not be used beyond $x < 0.1$. Yet in late 1960s this equation was introduced for calculating melt viscosity of a polyblend of LDPE lots (differing in MFI):

$$\ln (\eta) = \ln (\eta_i) + (1 - x) \ln (\eta_m) \tag{17}$$

without the composition limitation and with the viscosity, η_i substituted with Melt Flow Index (MFI). The same equation was again used for a description of the obvious multiphase melt of the polyblend polyethylene and polystyrene [9,28]; consequently, the synergism in melt viscosity was attributed to the scatter of experimental data. An important result of the research, the composition range in which there are particular rheological properties (shear viscosity synergism) was overlooked [9]. A similar deficiency in interpreting (Newtonian) melt viscosity-composition dependence for i-PP/HDPE systems [29] resulted from the use of Lee's mixing rule [22]:

$$\eta^a = \eta_i^a \cdot v_i + \eta_m^a \cdot v_m \tag{18}$$

where $v_{i,m}$ are volume fractions and a is an empirical exponent. Seventy years later the Lee's rule was again used in the form [29]:

$$\eta_0 = (w_i \cdot {}_i\eta_0^a + w_m \cdot {}_m\eta_0^a)^{1/a} \tag{19}$$

where η_0 are the Newtonian viscosities at 463 K (cone-plate measurements) and $w_{i,m}$ are the mass fractions of the components. Polyblends oriented application of Lee's rule was suggested by Dunlop. For the binary mixture of isotactic polypropylenes (i-PP) [30] Newtonian viscosity, η_0 is:

$$\eta_0 = w_i \left(\frac{\overline{M}_w}{_i\overline{M}_w} \right) \cdot \left(\frac{\overline{M}_w}{\overline{M}_n} \right)_i^{-1} \cdot {}_i\eta_0 + w_m \left(\frac{\overline{M}_w}{_m\overline{M}_w} \right) \cdot \left(\frac{\overline{M}_w}{\overline{M}_n} \right)_m^{-1} \cdot {}_m\eta_0 \tag{20}$$

where mol. mass: $_{i,m}\overline{M}_w$; $_{i,m}\overline{M}_n$ and viscosity $_{i,m}\eta_0$. The hyperbolic mixing rule of Bingham [13,31] in which the contributions of components are represented by fluidites ($\phi = 1/\eta$) rather than viscosities is:

$$\phi = {}_i\phi \cdot v_i + {}_m\phi \cdot v_m \tag{21}$$

where $v_{i,m}$ are volume fractions. Again, mixing rule in this form tends to suppress the indications of synergism [28].

One of the most confusing mixing rules for binary polyblends appears deceptively simple:

$$\eta^+ = \frac{\eta_i^+ \cdot \eta_m^+}{x \cdot \eta_i^+ + (1 - x)\eta_m^+} \tag{22}$$

where η^+ are "apparent" melt viscosities and x is mass fraction [32]. Because Equation (22) was derived for commercially attractive polyblends, and ostensibly from the generalized Maxwell model in the simple shear, substantial errors in the derivation and its dimensional inconsistency were neglected; moreover, it was extensively used [25]. It is evident that assumptions made in [32] are untenable:

- equal power law exponents for the components (almost Newtonian polyamide melt and distinctly pseudoplastic i-PP melt: $n_i = n_m = const!$)
- Uniform shear rates ($\dot{\gamma}$) in phases of distinctly different viscosities. These assumptions led to the derivation steps such as: $m_i \Sigma \dot{\gamma}_i = \gamma \Sigma_1 m_i$ (where m_i is the mass fraction of model elements) and $\dot{\gamma} = \gamma$ (shear strain!).
- power law in the form:

$$\eta^+ \cdot \dot{\gamma} = p_{12}^{n+} \tag{23}$$

where p_{12} = shear stress, the dimensions for "viscosity," η^+ involve the n^+ exponent. Rederiving Equation (22) in terms of the "conventional" power law:

$$p_{12} = p_{12}^0 \cdot (\dot{\gamma}/\dot{\gamma}^0)^n \tag{24a}$$

with $\dot{\gamma}^0 = 1 \; sec^{-1}$ or, using the standardized [17] viscosity, η^0:

$$\eta = \eta^0 \cdot (\dot{\gamma}/\dot{\gamma}^0)^{n-1} \tag{24b}$$

yields dimensionally correct viscosity yet complex, much less attractive mixing rule.

The first attempt to express composition dependence of the *rheological* properties of binary polyblends is in Bogue's [23] "quadratic" mixing ("blending") rule "BMEO":

$$H_1 = w_i^2 \cdot {}_iH_1 + 2w_iw_m \cdot {}_{im}H_1 + w_m^2 \cdot {}_mH_1 \tag{25}$$

in which the viscoelasticity of components and of the polyblend is represented by the first approximation relaxation time spectra ${}_iH_1$ and ${}_mH_1$ and components interaction, though unspecified, is embodied in an "interaction spectrum," ${}_{im}H_1$ term. Subsequently [24] Bogue and Racin developed a unique approach to binary polyblends rheology-composition dependence in formulating a constitutive equation for these systems—an equation in which interaction is embodied in the "BMEO manner," i.e., in the "mixed" relaxation time, τ_{pq}^+ [24]. The constitutive equation used in the simplified "binary" form is:

$$p_{ij} = -p\delta{ij} + G^+ \cdot \int_{-\infty}^{\tau} \sum_{p=1}^{N} \cdot \sum_{q=1}^{N} w_p w_q \frac{\exp\left(-\dfrac{t-t'}{\tau_{pq}^+}\right)}{\tau_{pq}^+} \cdot C_{ij}(t,t')dt' \tag{26}$$

where

p_{ij} = stress tensor
c_{ij}^{-1} = Finger rate of strain tensor
G^+ = a constant having dimensions of modulus
$w_{p,q}$ = mass fractions of p and q, the component—polymers

Important interaction parameter τ_{pq}^+, with $+$ designation stipulating "feature characteristic for the total (or overall) for the polyblend," is:

$$\tau_{pq}^+ = \frac{\sqrt{T_p \cdot T_q}}{1 + a\sqrt{T_p T_q} \cdot \mathrm{II}_d} \tag{27}$$

where

$\mathrm{II}_d^{1/2}$ = time averaged 2nd invariant of the rate of deformation tensor
$T_{p,q}$ = overall (characteristic) relaxation times for the p and q polymers
a = an adjustable ($a = 0.3$ to 0.8), dimensionless constant

Characteristic relation times, e.g., τ_p are related to molecular mass via a "reference" relaxation time:

$$\frac{\tau_p}{\tau_{ref}} = \left(\frac{\overline{M}_p}{M_{ref}}\right)^2 \cdot \left(\frac{M^+}{M_{ref}}\right)^{1.5} \tag{28}$$

where τ_{ref} is a characteristic relaxation time observed for a polymer of molecular mass, M_{ref}. Bogue's approach, therefore, allows to derive material functions for the binary polyblends in which the interaction of components is explicitly accounted for. For example, the first normal stresses difference, N_1 is:

$$N_1 = \frac{2(p_{12})^2}{G} \cdot \frac{W_i \dfrac{\tau_i}{1 + a\dot{\gamma}\tau_i} + W_m \dfrac{\tau_m}{1 + a\dot{\gamma}\tau_m}}{W_i \sqrt{\dfrac{\tau_i}{1 + a\dot{\gamma}\tau_i}} + W_m \sqrt{\dfrac{\tau_m}{1 + a\dot{\gamma}\tau_m}}} \tag{29}$$

where

p_{12} = shear stress
G = characteristic modulus
$w_{i,m}$ = mass fractions

The phenomenological approach represents composition dependence of melt *rheology* for binary polyblends with an unspecified and probably weak interaction and its ability to predict particular rheology compositions [3] has yet to be evaluated.

Mixing rules discussed so far "suppressed" manifestation of synergism by their exponential form. Certain progress however is evident in *predicting shear viscosity* of the binary polyblends in the entire range of its composition, i.e., including synergistic compositions. The progress is mostly due to Crossan-Carley (CC) research on the components interaction effects [25] and to studies of Kasajima and coworkers (KM) [26] on the shear rate dependence of the melt viscosity-composition relationship. Kasajima et al. proposed a "mixing effectiveness indicator" (MEI) displaying "interaction and goodness of mixing" effects. Although KM suffers from (untenable) assumption of uniform shear rate field in a molten, flowing polyblend it appears correct conceptually in attempting to account for the shear dependence of viscosity. In the CC approach the interactions on both macromolecular ($M_{i,m}$) and supramolecular (specific melt volume, v_T) levels are accounted for. It seems to offer therefore a satisfactory description for most of the binary polyblends studied. Success of the "CC" mixing law is due to the fact that the components viscosities used are kinematic, x, i.e., the *product* of dynamic viscosity, $\eta_{i,m}$ and specific melt volume $_{i,m}v_T$. Specific volume appears to be associated with the phase structure (morphology) of the molten polyblend [5,34,49]:

$$x_{1,2} = \eta_{i,m} \cdot {}_{i,m}v_T \tag{30}$$

The CC mixing rule for the (shear) viscosity is:

$$x = x_i^3 \cdot \ln x_i + x_m^3 \cdot \ln x_m + 3 x_i^2 x_m \cdot \ln x_{im} + 3 x_i x^2 \ln x_{mi}$$

$$+ 3 x_i^2 x_m \left(\frac{2M_i + M_m}{3} \right) + 3 x_i x_m^2 \ln \left(\frac{M_i + 2M_m}{3} \right) \tag{31}$$

$$+ x_i^3 \ln M_i + x_m^3 \ln M_m - \ln (x_i M_i + x_m M_m)$$

where

$x_{i,m}$ = molar fractions
$M_{i,m}$ = molecular mass of the components

Carley also demonstrated the satisfactory (predictive) performance of a much simpler, empirical mixing rule:

$$\ln \eta = A w_i^3 + B w_i^3 + C w_i^2 w_m + D w_i w_m^2 \tag{32}$$

where

A,B = $\ln \eta_i$, $\ln \eta_m$, respectively
C,D = $3 \ln \eta_{im}$, $3 \ln \eta_{mi}$, respectively, [25b]

Since Equation (32) offers little advantage over (31) with respect to accounting for *both* interaction of components *and* shear stress level effects on viscosity-composition relationships, a closer evaluation of KM approach seems to be justified.

Basic merit of KM is in that it addresses *both* the composition and shear dependence of the polyblend viscosity. Albeit the assumption of a uniform shear rate within the polyblend melt is untenable [15,16], the concept of expressing shear dependence via its effect on composition is viable. Nominal composition, expressed here in mass fraction of the higher viscosity component, c is converted to the shear interaction corrected composition (or the "mixing effectiveness indicator," MEI):

$$C^+ = \Phi(C,\dot{\gamma},T) \cdot c \tag{33}$$

where the correction factor $\Phi = \Sigma \{\lambda_j(c,T)\} \cdot (\ln \dot{\gamma})_j$, where $\dot{\gamma}$ is the shear rate and T is the temperature, could also be evaluated in terms of "component shear stress," i.e., the shear stress p_{12} induced in component polymer phases by the "overall" shear rate $\dot{\gamma}$. KM approach *incorrectly* [15] assumes that resultant shear stress, p_{12} in a polyblend is a geometric mean of stresses in the components:

$$p_{12} = {_i}p_{12}^{C^+} \cdot {_m}p_{12}^{(-C^+)} \tag{34}$$

and, therefore, that the correction factor for effective composition could be evaluated from:

$$\Phi(c,T,\dot{\gamma}) = -\frac{1}{C} \cdot \frac{\ln p_{12} - \ln {_m}p_{12}}{\ln {_i}p_{12} - \ln {_m}p_{12}} = \frac{1}{C}\left[\ln \frac{\eta^0 \cdot \dot{\gamma}^n}{\eta_m^0 \cdot \dot{\gamma}^{nm}} \cdot \ln \frac{\eta_i^0 \cdot \dot{\gamma}^{ni}}{\eta_m^0 \cdot \dot{\gamma}^{nm}}\right]^{-1} \tag{35}$$

where n, η^0 are power law ($p_{12} = \eta^0 \cdot \dot{\gamma}^n$) parameters, η^0 is the standardized viscosity and n is the power law exponent [17]. Data on HDPE (polystyrene system) indicate that the deviation from one to one correspondence between the nominal (C) and an "effective" (C^+) concentrations decreases with increasing shear rate, $\dot{\gamma}$ and, moreover, that the deviation is most pronounced at a composition range $(C = 0.75$ for the system in consideration) in which $C^+ \cong 1.5C$. Consequently, the MEI factor is a measure of the degree of synergism, and may be expressed by:

$$\ln a_s = [1 - \Phi(c,\dot{\gamma},T)] \cdot c \cdot \ln \frac{{_m}p_{12}(\dot{\gamma},T)}{{_i}p_{12}(\dot{\gamma},T)} \tag{36}$$

In summary, the KM approach reformulated in terms of components' ("inherent") shear rates, rather than stresses [15,16]—RKM offers an effective route for engineering design of binary, microheterogeneous polyblends with respect to melt viscosity. On the basis of available, verified data for number of binary polyblends [3,25,33] the effective composition approach (with shear stress as the correcting parameter) could be extended to polyblends melt elasticity (i.e., first normal stresses difference, N_1), elongational viscosity (e.g., estimated from Cogswell convergent flow [4a]) and to effective relaxation time (see Figure 16). Considering that two adjustable constants are required for the "mixing rule" constitutive equation [24], the "effective composition" (RKM) approach to the rheological design of binary polyblends appears attractive.

Tentative Evaluation of Viscosity Predictive Models for the Polyblends

The evaluation was performed [35] using melt rheology data presented in the standardized form for HDPE/LDPE polyblend d, for which both melt specific volume and mol. mass as function of composition are available [3a]. In the evaluation only basic assumption of the actual concentration C^+ dependence on the nominal concentration, C and on the flow induced "compatibilization" was preserved. The later (interaction) is expressed by the correction factor, Φ given by Equation (35). The "Arrhenius averaging"—Equation (16) was employed to the apparent shear rate, $\dot{V}$. Such a procedure is justifiable in terms of contemporary understanding of the flow in multiphase, viscoelastic systems [21]. Equation (34) is replaced, therefore by:

$$\dot{V} = \dot{V}_i^{C^+} \cdot \dot{V}_m^{(1-C^+)} \tag{37}$$

Using power law in a simplified form: $p_{12} = \eta^0 \cdot \dot{V}^n$ ($\dot{V}$ = apparent shear rate) and designating again $(p_{12}/\eta^0) = S$, and $1/n = q$, we have, at a p_{12} level *equal* in *both* phases:

$$\ln S^q = C \cdot \Phi \cdot \ln S_i^{q_i} + (1 - C\Phi) \cdot \ln S_m^{q_m} \tag{38}$$

and

$$\Phi = \frac{q \cdot \ln S - q_m \cdot \ln S_m}{C\,(q_i \cdot \ln S_i - q_m \cdot \ln S_m)}$$

$$= \frac{q \cdot q_m\,(n_m - n_i) \cdot \ln p_{12} + \ln \{(\eta_m^0)^{q_m} \cdot (\eta^0)^{-q}\}}{C\,\{q_i \cdot q_m\,(n_m - n_i) \cdot \ln p_{12} + \ln [(\eta_m^0)^{q_m} \cdot (\eta_i^0)^{-q_i}]\}} \tag{39}$$

obviously S, S_i and S_m differ because η^0, η_i and η_m^0 are a priori different.

From the analysis of data it follows that the correction factor Φ is stress dependent (Figure 17) and that it reflects complexity of composition-shear stress interaction (Figure 18). The plot of the effective (C^+) versus the nominal composition, displays combined composition and stress level dependence of the components' interaction. The RKM model is then a qualitative guide for selecting composition of a polyblend. Hence, in locating the particular rheology compositions and using the standardized melt viscoelasticity functions in conjunction with the discrete composition sampling [3a] procedure one may test the correctness of the composition selection with respect to the polyblend processability using RKM.

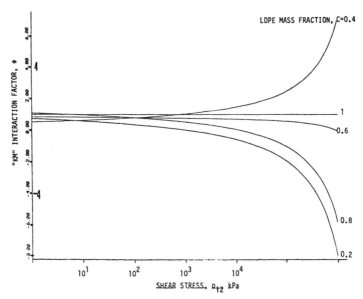

FIGURE 17. Shear stress dependence of the interaction factor, Φ for the "d" polyblends at the nominal compositions, c of 0.2, 0.4, 0.6 and 0.8 (wt. fraction) of LDPE [3a, 35]—B. H. Luor.

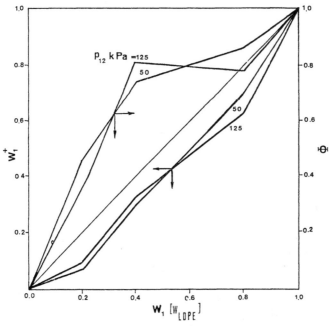

FIGURE 18. Composition dependence of the interaction factor, Φ and the effective compositon, C^+ at various shear stress p_{12} levels (courtesy B. H. Luor [35]). "d" polyblend at 463 K [3a].

All the merits notwithstanding, RKM is oversimplified because of the fundamental assumption of the Arrhenius (first order) interaction. In this respect the "CC" model Equation (31) is more realistic; it accounts also for rheological, macromolecular and supramolecular interactions of 2nd order. The performance of CC was verified again using data for the d polyblend; the plot of the stress dependent "2nd interaction" viscosity x_{21} is given in Figure 19. Comparison of the performance of RKM and CC models with that of Hayashida's model is displayed in Figure 20 for HDPE/LDPE polyblend.

Appendix A: Elasticity Ratio in Binary Polyblends— The Normal Stresses Ratio $(N_1)_r$ Form

Elasticity, ψ_1, considered here as the criterion for the shape of minor phase domains [15] is usually defined by: $\psi_1 = N_1/\dot{V}^2$ where $\dot{V}$ is the apparent shear rate. It follows, therefore, that the elasticity ratio $(\psi_i)_r$ will depend, at a shear stress level much more on the ratio of shear rates $(\dot{V}_i/\dot{V}_m)^2$ than on the ratio of normal stresses $(N_1)_r$.

It is of interest, therefore, to compare the function of normal stress difference ratio, $(N_1)_r = f(p_{12})$ with the $(\psi_1)_r = f(p_{12})$ function. Assuming power law dependence of N_1 we have [3]:

$$N_1 = N_1^0 \cdot \dot{V}^m = N_1^0 \cdot \left(\frac{p_{12}}{\eta^0}\right)^{\frac{m}{n}} \tag{A.1}$$

putting $m/n = z$ we have:

$$(N_1)_r = ({}_iN_1/{}_mN_1) \cdot (N_1^0)_r \cdot (\eta_1^0)^{zi} \cdot p_{12}(z_1 - z_m) \tag{A.2}$$

and

$$(N_1)_r = \chi \cdot p_{12}^e \tag{A.3}$$

where

i = minor phase
m = matrix
p_{12} = shear stress
$\Delta\dot{V}$ = "inherent" shear rate difference, $\Delta\dot{V} = \dot{V}_i - \dot{V}_m = 0$
n_i, n_m = power law exponents
q_i, q_m = the reciprocals of n_i, n_m, respectively

The VanOene criterion for the transition in phases' distribution $\Delta N_1 = N_{1r} - N_{1m}$ could be directly expressed by $(N_1)_r$:

$$\Delta N_1 = (N_1)_m \cdot [(N_1)_r - 1] \tag{A.4}$$

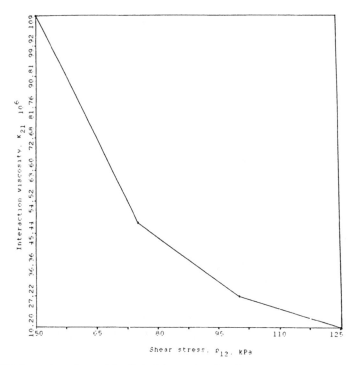

FIGURE 19. Stress dependence of the Carley's interaction viscosity, x_{21}: trendline simulation for "d" polyblend, 463 K (courtesy B. H. Luor [35]).

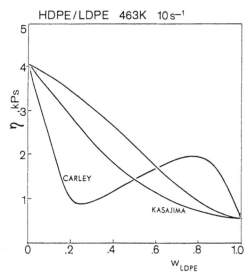

FIGURE 20. Comparison of Carley, Kasajima and Hayashida mixing rules for shear viscosity and HDPE/LDPE ("d") polyblend at 463 K [35].

Relationships between the parameters of $(\psi_1)_r$ and $(N_1)_r$ functions are:

$$\chi' = \chi \cdot (\eta_i^0)^{2q_i} \cdot (\eta_m^0)^{-2q_m} \tag{A.5}$$

and

$$e' = e - 2(q_m - q_i) \tag{A.6}$$

[compare Equation (10)]. For a realistic set of $\eta_{i,m}^0$ and $n_{i,m}$ values one could estimate $\chi'/\chi \cong 10^{-10}$ and $e'/e \cong 6$. Hence the predictions of a binary polyblend morphology based on $(\psi_1)_r$ and $(N_i)_r$ functions may differ significantly.

Appendix B: Emulation of the Polyblend Morphology—Part I

ALGORITHM FOR ENGINEERING APPROXIMATIONS: DISPERSE MORPHOLOGY, DOMAIN LEVEL

Viscosity and elasticity functions of the component polymers expressed as (stress dependent) ratios VR (η_r) and ER (N_{1r}). Using trucated log parabola (power law) or log parabola approximations for the (V, E) R functions one has:

$$(VR)_1 = \eta_r^0 \cdot \frac{p_{12}^{qm}(\eta^0)^{-qm} + \Delta\dot{V}}{p_{o12}^{(1-q_i)} \cdot (\eta_m^0)^{(qm-1)}} \tag{B.1}$$

$$(VR)_2 = \eta_r^0 \cdot \frac{p_{o12}^{(1-q_i)} \cdot (p_{12}^{qm})^{(q_i-1)}}{[p_{12}^{q_i} \cdot (\eta_i^0)^{-q_i} \cdot \Delta'\dot{V}]^{(nm-1)}} \tag{B.2}$$

where 0 superscript designates "standard" shear rate of 1 sec^{-1} [17] $\dot{V}_{i,m}$—inherent shear rates and $\Delta'\dot{V} = \dot{V}_m - \dot{V}_i \geq 0$

$$(VR)_3 = (\eta_i^0)^{q_i} \cdot (\eta_m^0)^{qm} \cdot (p_{12})^{(qm-q_i)} \tag{B.3}$$

and, finally:

$$\log (VR)_4 = (a_i - a_m) + (b_i - b_m) \log p_{12} + (c_i - c_m) \log^2 p_{12} \tag{B.4}$$

where a, b and c are parameters of the viscosity function approximation of the form:

$$\log (\eta) = a + b \log p_{12} + c \log^2 p_{12}$$

The elasticity ratio, ER is:

$$N_{1r} = \frac{N_{1r}^0}{(\eta_i^0)^{z_i}} \cdot (\eta_m^0)^{z_m} \cdot p_{12}^{(z_i-z_m)} = N_{1i}/N_{1m} \tag{B.5}$$

where

$$z = m \cdot q$$
$$m = \text{elasticity exponent}$$

Morphology type—disperse (drops, cylinders or ribbons of the minor phase) or stratified ("sandwich") is estimated from components elasticity, N_1 and interfacial tension, v. As a rule of thumb one may assume that for $\Delta N_1 (= N_{1i} - N_{1m}) > 0$ the disperse morphology prevails [15].

Stress dependent domain size, $<d_i>$, is then estimated from [see Equation (15)]:

$$<d_i> = \frac{(\eta^0)^{-2q} \cdot 40 \, N_1 \cdot v \cdot (\eta_r + 1)^2}{(19\eta_r + 16)^2 \cdot \eta_i^2 \cdot p_{i\dot{1}}^{2q} \cdot v_i} \tag{B.6}$$

where v_i is the volume fraction of the minor phase, v—interfacial tension. Equation (B.6) is derived from constitutive equation for a simple emulsion model [20]; the "nonindexed" parameters refer to the polyblend.

From the above a number of derived characteristics could be estimated. These are pertinent for design of the extrusion compounding process; yielding estimates for e.g., shear rate at the domain rupture, $\dot{V}_i$:

$$(\dot{V})_i = \frac{v}{2\eta_r <d_i>} \cdot \frac{(16\eta_r + 16)}{(19\eta_r + 16)} \tag{B.7}$$

or the *circulation time*, t_c, in a domain:

$$t_c = (4\Pi_d / \dot{V}_m) \cdot (\eta_r + 1) \cdot [\eta_r(\eta_r + 2)]^{-1/2} \tag{B.8}$$

($\dot{V}_m$ is the apparent shear rate in the matrix) are available [51]. It should be underscored that VR used here is function of stress [see Equations (B.1–B.4)].

Appendix C: Prediction of Morphology—Part II: PROCEDURE

Component and composition selection for a microheterogenous POLY-BLEND/ALLOY.

Input:

component polymers' shear and elongation viscosity functions, melt elasticity expressed as the stress dependent material functions; specific melt volume and surface tension, their temperature and pressure dependence; phase diagram and the free energy of mixing; performance characteristic(s) of the components intended for engineering polyblends/alloys; crystallization and spherulite formation kinetics; macromolecular structure: M, MWD.

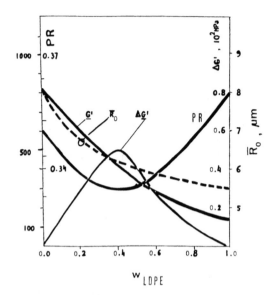

FIGURE 21. Composition dependence of: G' = the shear modulus (Faucher/Kerner model [46]); $\Delta G'$ = difference of predicted and experimental moduli, Poisson ratio, æ and the mean spherulite radius, $\bar{R}^o$ for "d" polyblend at 293 K [35].

Components' selection:
   analysis of the viscosity ratio, *VR* and of the elasticity ratio (principal normal stress difference, N_1), *ER* functions at a selected inherent shear rate (IRS) difference [Equations (B.1–B.5)]

   ⟶ components' selection

   ⟶ the type of phase structure (morphology) at a selected major component and the domain size, $<d>$ [Equation (8)]

   ⟶ estimation of the (shear stress dependent) size of the minor phase domain at a selected composition; verification of the type of the morphology [Equations (B.6) and (15)]

Composition selection:

   (*) = emulation with the "mixing laws"

 * location of the composition of the (melt) viscoelasticity synergism at a selected level of (the overall) shear rate ["CC" —Equation (31)]

   ↓

 * evaluation of the stress dependence of the synergism ["RKM"—Equation (36)]

   ↓

* evaluation of a performance characteristic with the modulus—composition dependence from, e.g., Kerner/Faucher model ([46], Figure 22)

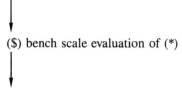

($) bench scale evaluation of (*)

$ select:
  - mixer/mixing process parameters/stress level/duration, total strain, stress/temperature field
    (experiment variables for the mixer [48,79, 82])
  - the component-polymers with respect to the limits of viscosity and elasticity ratio functions: max, min, and the median of the range
  - the interface modifying additive (compatibilizer): yes/no
  - the compositions (minimum); the selected, complementary and 1:1

$ prepare the polyblends/alloys and evaluate their melt rheology and concomitant morphology (including interface) (Figure 22)
    (final verification of the polyblend design)
  - pattern of the melt flow: viscometric/elongational
  - geometry of the melt flow: confined/free (film!)
  - stress levels within the range encompassing the intended processing
  - phase structure following general route

$ verify the processability and performance of the polyblend/alloy in comparison with these in the "design" stage and proceed with the technical evaluation (pilot plant scale, industrial processing)

MIXING/COMPOUNDING POLYMERS INTO POLYBLENDS/ALLOYS [1a,17,29b,43,54,65,68,69,81]

Although the morphology of polyblends/alloys depends ultimately on the viscoelasticity ratio functions and the interface, it is essential to realize that the morphology and the properties are controlled by *two* modes of mixing. The *dispersive* (DIS) mixing [65] leading to a selected size and shape of the minor phase domains (particles of the minor phase) is the process over which a control via the above discussed procedures is feasible. The control over the spatial distribution of the domains resulting from the *distributive* (DIT) mixing [17,68,69,81] is much weaker because of notoriously high melt viscosity and elasticity, which combined with very slow diffusion strongly attenuate the mass transport. In effect the successful design of an industrial polyblend has to accommodate a number of experimental steps bridging the gap in understanding the mechanism of "DIT" in the variety of the polyblends fabrication processes listed in Figure 23. A comprehensive description of the morphology of an industrial polyblend in terms of the "mixing quality" measures [68,3b–c,35] is attempted in Figure 24 [57]. Types of the morphology are illustrated in Figure 25 (top row) which combines these with the convincing demonstration of the mixing process effects on the composition dependence of melt viscosity in the polyisoprene/polybutadiene blend. The viscosities measured at the temperatures [°C] indicated are given for the polyblend prepared (fabricated) by: 1—twin roll mill mixing, 2,5—dissolution with subsequent evaporation of the solvent, 3—coprecipitation from a "joint" solution, curve 4 reflects the effect of MWD change (components and fabrication—same as 5) [29b]. In the design of a polyblend with an account for the mixing process the interrelation diagram (Figure 26) is useful. After selecting the components from the variety of grades of the component-polymers on the basis of a desired combination of the viscosity and elasticity ratio functions (the ratios may be both high—H:H, both medium—M:M, both low—L:L, etc.) the composition is selected, primarily in the consideration of the composition synergism [3a]. Each of the selected compositions should be evaluated in comparison to the components (with the same mixing/processing history), the complementary and 1:1 compositions. At this stage the option of the interface modification (compatibilizer and/or a reactive reinforcement) should be considered. In selecting a mixing process at least four options should be considered: continuous and batch variants of the distributive (DIT) and of the dispersive (DIS) mixing processes. Obviously the actual number of options is much larger with the industrial mixers: each of these represent a combination of DIT/DIS functions and can be operated in a range of the stress field/characteristic time levels.

It is essential, therefore, to estimate these for the mixer(s) selected, e.g., by using the generalized Newtonian approximation for the flow field in the mixer(s) combined with the power law rendering of the polyblend flow properties. An illustration of the procedure introduced by Tadmor is given in Figure 27: for the essential parts of the industrial mixers the estimates are given for the characteristic time and for the shear stress level. In turn stress level is used for estimating the outcome of DIS mixing by calculating the mean domain size, $<d>$, the main characteristic of the morphology via the SIMULBLEND procedure outlined in Figure 28 and discussed in Appendix C—Part 2. A comparison of the predicted and measured on the "bulk" specimen of a model polyblend (LDPE/PS) fabricated in the predominantly DIT (industrial) mixer is given in Figure 29. The actual measurements of the domain size are currently performed with the image analysers [3c,8l] which digitize microphotographs (count the number of pixels encased within the boundary of the domain or measure length of lines crossing over the domain images—Figure 30) and present the statistics of the morphology features. Alternatively an estimate of the domain size may be retrieved from measurements of the (optical) turbidity of the semimolten polyblend (Figure 31). A comprehensive analysis of the polyblend/alloy morphology and its dependence on melt flow requires at least two specimens to be analysed: Bulk (b) is a specimen collected from within the mixer; the "g" specimen represents the typical form of an industrial polyblend/alloy, i.e., pellets. The time dependence and flow

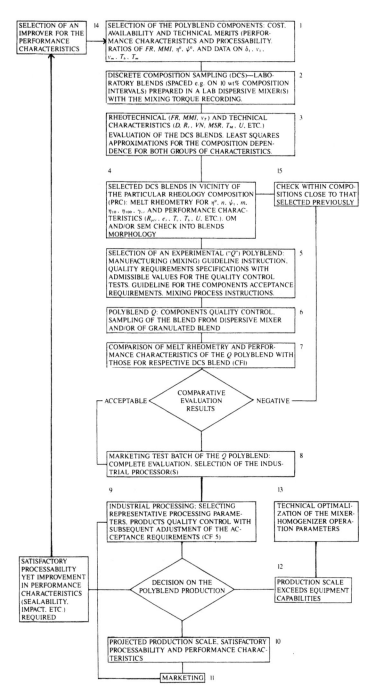

FIGURE 22. Application of melt rheometry in selection of components and composition in the development of polyblends.

195

d_{23}^4 = density at 23°C
DCS = discrete composition sampling (blend)
e_r = elongation at break
FR = flow rate (melt—ASTM-D1238)
m = elasticity exponent ($\psi_1 = \psi_1^0 \cdot \dot{\gamma}^{m-2}$)
MMI = melt memory index[1]
MSR = melt stability ratio $\left(\dfrac{FR\ \text{after 30 mins}}{FR\ \text{after 6 mins}} \right)$
n = viscosity exponent ($\eta = \eta^0 \cdot \dot{\gamma}^{n-1}$)
OM = optical microscopy
PRC = particular rheology composition
R = tensile yield stress
R_r = tensile strength
SEM = scanning electron microscopy

T_B = brittleness temperature
T_m = melting (softening) temperature
T_v = softening temperature after Vicat
U = mixing energy
v_T = specific melt volume @ temperature T
VN = viscosity (soln.) number
$\dot{\gamma}_{cr}$ = critical shear rate[2]
δ = solubility parameter
η^0 = melt viscosity @ 1 sec^{-1}
$\eta_{10},\ \eta_{100}$ = viscosity of melt @ 10 and 100s^{-1}, respectively
v_i = (melt) surface tension
ψ_1^0 = melt elasticity @ 1 sec^{-1}

*For three or more components polyblend the statistical design experiments [48,79,83] should be followed.
**Rational approach requires parallel DCS blends preparation with at least two types of the mixer (mostly DIS —e.g., DBM, Figure 23 and mostly DIT—e.g., static mixer).
[1]Extrudate swelling measured on "as received" FR extrudate.
[2]At which severe surface distortion is observed on extrudates.

FIGURE 22 (continued). Application of melt rheometry in selection of components and composition in the development of polyblends.

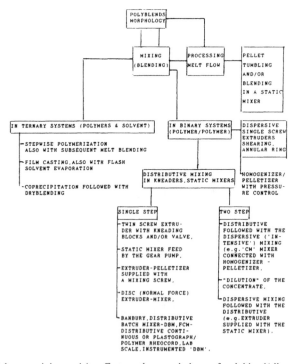

FIGURE 23. Polymer mixing and its effect on the morphology of polyblend/alloy: a classification scheme.

Type of the phase morphology	Description	Experimental Data	Related feature of the microstructure (phase morphology)
	Composition (concentration) uniformity	Mean concentration: $\bar{a}$, $\bar{b}$	Probability of a chance (not related to mixing) deviation from a nominal concentration α
		Variance of the concentration distribution*	Degree of uniformity. M (Lacey index) [Equation (9)]
Disperse	DISPERSION: FINENESS	Mean size of the minor phase domain. $<d>$	Degree of dispersion (D_d) [Equation (10)]
	(local microstructure from e.g. high mag. microscopy)	Variance of domain size distribution, S_d^2	Uniformity of dispersion (D_{S_d}) [Equation (11)]
Stratified and/or interpenetrating		Mean transversal length: $\bar{T} = \bar{T}_a + \bar{T}_b$ variances of transversal ("dipole") length: $S_{T_a}^2$, $S_{T_b}^2$	Degree of dispersion (D_l) [Equation (12)] where l="dipole" length Uniformity of dispersion: D_{sa}, D_{sb} [Equation (13)]
General (TEXTURE)	DISPERSION: MODE (texture)	Spatial distribution of concentration	Concentration correlation function. $k(r)$*
		Spatial distribution of domain size or of transversal lengths	Size correlation function: [Equation (14)] $<d>$', "-domain sizes at regions separated by the distance r*

Auxiliary parameters characterizing TEXTURE [68] for both size and concentration distributions. S_l = linear scale of correlation (LSC): Equation (15); S_v = volumetric scale of correlation (VSC): Equation (16); where r = the distance at which $K(r) = 0$.
*See e.g. References [57,68,81].

FIGURE 24. Comprehensive description of polyblend/alloy "mixedness:" intensity, scale of segregation and texture (TEXTURE is understood here as long range, or supradomain uniformity).

Polyblends: Texture and Morphology Descriptors

Equations

Lacey index (Schenkel):

$$M = (S_0^2 - S_u^2)/(S_0^2 - \underline{S}^2) \tag{9}$$

Degree of dispersion:

$$D_d = [<d_0> - <d>]/<d_0> \tag{10}$$

Uniformity of dispersion:

$$D_{sd} = S_d^2/<d> \tag{11}$$

Degree of dispersion:

$$D_1 = (l_0 - <l>)/l_0 \tag{12}$$

Uniformities of dispersion:

$$\left. \begin{aligned} D_{sa} &= S_{l_a}^2/<l_a> \\ D_{sb} &= S_{l_b}^2/<l_b> \end{aligned} \right\} \tag{13}$$

Size correlation function:

$$K_d(r) = 1/n \cdot S_d^2 \sum_1^n (<d'> - <d>) \cdot (<d''> - <d>) \tag{14}$$

Linear scale of correlation (LSC):

$$S_l = \int_0^{r_m} K(r)\, dr \tag{15}$$

Volumetric scale of correlation (VSC):

$$S_v = 2\pi \int_0^{r_m} K(r)\, dr \tag{16}$$

$\underline{S}^2$ = variance in a perfectly mixed state: $\underline{S}^2 = p(1 - p)v_e/V$

S_0^2 = variance in a perfectly unmixed state: $S_0^2 = p(1 - p)$

l = "dipole" length (IS ferret)

$<d',''>$ = domain sizes at the regions separated by distance r

r_m = distance @ which $K(r) = 0$

198

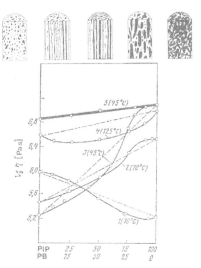

FIGURE 25. Main types of polyblends' morphology and their dependence on the preparation technique. See text and [29b] for description of the viscosity-composition dependence.

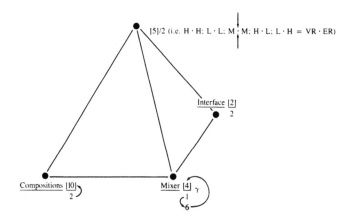

Key:

$\frac{[M]}{N}$ = [*minimum number of variable levels for comprehensive program*]/levels selected for the program

$(VE)_r$ = ratio of viscosity, η_r, and on elasticity (N_1), functions (p_{12})

(%) = composition (minor phase content), e.g. 0, 33, 50, 67 and 100 wt.%,

γ = total strain in mixing [*distributive: dynamic* (3 variants) or *static* mixer, *dispersive* (laminar-grain size, TRD, γ = total strain to be specified) at a single level of process parameters, *coprecipitation*].

FIGURE 26.

199

TIME OF RESIDENCE:

$$\text{SCREW: } \langle t \rangle = \dot{V}\phi/Q = \dot{V}/Q_d\,[2F_d \cdot H/W(1 + F_d\,H/\phi_w)^{-1}]$$

$$\text{DISC: } \langle t \rangle = \langle t \rangle_{screw} \times 2$$

KNEADING BLOCKS:

$$\langle t \rangle = \dot{V}/q(L + 1/\dot{V}_0)1/(1 + k_0)\cdot(1 + z_0)/2 \cdot 1/(1 + k_0 z_0)\cdot[1 + 2k_0)/z_0]\{1 + z_0^2\,k_0(2/1 + 3z_0)\}$$

ϕ = fill factor; $\dot{V}$ = velocity; $\dot{Q}$ = flow rate; F_d = drag flow factor; H/W = channel shape.

STRESS (shear), p_{12}:

$$\text{SCREW: } \langle p_{12} \rangle_{L,U} = p_{12}^0 \cdot \langle \dot{\gamma} \rangle_{L,U^n}$$

$$p_{12} = p_{12}^0\,\dot{\gamma}^n$$

KNEADING BLOCKS:

$$\langle \dot{\gamma} \rangle_{KB} = (\dot{V}_0/H_0)\cdot(1 + z_0)/2 \cdot [1 + z_0^2\,k_0(2/1 + z_0)]\,(1 + z_0 k_0)^{-1}$$

$$\langle p_{12} \rangle_{KB} = p_{12}^0\,\dot{\gamma}\,KB^a$$

$\underline{n}$ = mean exponent; $z_0 = H_0/H_1$; $k_0 = 1/L$; V_0 = velocity.

SIMILAR ESTIMATES FOR:

$\langle \gamma \rangle$ mean strain
$\langle \dot{\epsilon} \rangle$ rate of stretching
$\langle \epsilon \rangle$ stretch ratio

Example estimate for E mixer in #5491 experiment [J. E. Curry]: $\langle p_{12} \rangle_{KB} = 0.25 - 2.0$ MPa [80].

FIGURE 27. Estimates for process parameters for domains' formation in fabrication of polyblends in extrusion (Z. Tadmor, 1986).

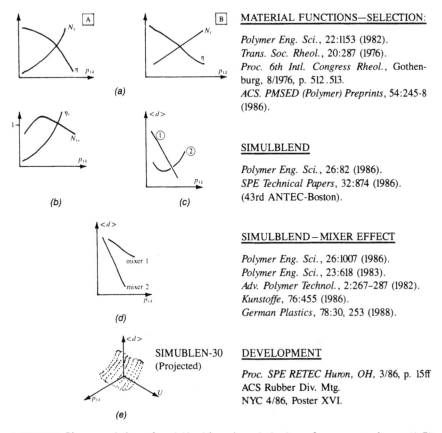

MATERIAL FUNCTIONS—SELECTION:

Polymer Eng. Sci., 22:1153 (1982).
Trans. Soc. Rheol., 20:287 (1976).
Proc. 6th Intl. Congress Rheol., Gothenburg, 8/1976, p. 512.513.
ACS. PMSED (Polymer) Preprints, 54:245-8 (1986).

SIMULBLEND

Polymer Eng. Sci., 26:82 (1986).
SPE Technical Papers, 32:874 (1986).
(43rd ANTEC-Boston).

SIMULBLEND—MIXER EFFECT

Polymer Eng. Sci., 26:1007 (1986).
Polymer Eng. Sci., 23:618 (1983).
Adv. Polymer Technol., 2:267–287 (1982).
Kunstoffe, 76:455 (1986).
German Plastics, 78:30, 253 (1988).

SIMUBLEN-30
(Projected)

DEVELOPMENT

Proc. SPE RETEC Huron, OH, 3/86, p. 15ff
ACS Rubber Div. Mtg.
NYC 4/86, Poster XVI.

FIGURE 28. Phase morphology of a polyblend from the melt rheology of component-polymers (A,B) "SIMULBLEND" with references: (a) shear stress dependent material functions, (b) Viscosity and Elasticity Ratio functions, (c) two patterns of the SIMULBLEND generated domain size $<d>$ = stress relationship, (d) minor phase domain size $<d>$ = stress dependence: anticipated effect of the mixing history, (e) complete SIMULBLEN-3D expected to yield the domain size $<d>$ as a function of stress (p_{12}) and a mixing process parameter, e.g., energy input (U) [49].

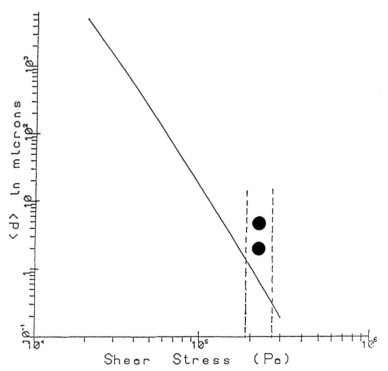

FIGURE 29. Domain size prediction with SIMUBLEND • (courtesy: S. D. Dagli).

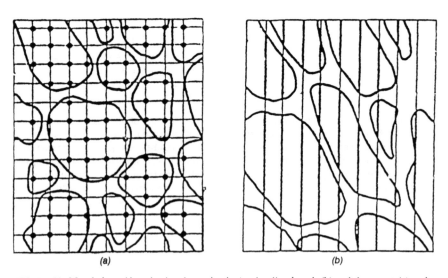

FIGURE 30. Morphology (domain size determination) using line length (b) and dot count (a) techniques [29b]. Image analyzers typically rely on the (a) method.

202

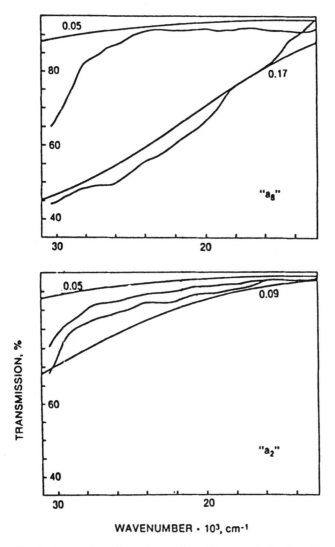

FIGURE 31. Complementary phase size evaluation from light transmission through semi-molten samples of "a_8" and "a_2" i-PP/LDPE polyblends. Smooth solid lines—theoretical transmission curves for suspensions of opaque solids, the particle (i-PP, minor phase) size (diameter annotated in mm). Experimental transmission traces indicate size of i-PP domains ("particles") [35,55].

segregation inherent to the melt flow imposed morphology of the polyblends makes this extensive characterization necessary.

In the industrial practice the "design" of a polyblend/alloy is much more complex, mainly because of the batch-to-batch variation in properties of the component polymers, diversity of the processing operations and, finally the time constraints of the development research. It is therefore convenient to adopt a more empirical procedure of the *discrete composition sampling* (DCS) in which, for the selected components, specimens of the polyblend are prepared on the bench-top scale at the composition increments of 5–10wt%. The specimens are subsequently characterized in a sequence.

COMPUTER PROGRAMS IN THE ENGINEERING OF THE POLYBLENDS/ALLOYS—AN OVERVIEW

Complexity of rheology, physics and chemistry of polymers, further compounded by poor understanding of the distributive mixing and of the interfacial phenomena prevented more comprehensive design attempts involving "mixing rules" [3c]. A progress was feasible only after the high speed/large memory computers became available.

Current programs for the design of polyblends/alloys constitute two groups:

- The structure-matching (thermodynamics of phase equilibria and solubility enhancement): these are typically proprietary, developed and operated within the main resin manufacturing companies (e.g., Dow, Celanese, General Electric) interactive graphics software packages. The output consists mainly of a guide for an adjustment in the component-polymers structure: MWD, branching and/or incorporation of active groups.
- The performance/processability targeted selection of the components, composition with an account for the mixing process effects. These programs, especially in the case of binary mixtures of the commercial grade polymer-components, might be represented by SIMULBLEND [3c], combined with the DCS procedure. Mixing process is characterized in the factorial design experiments [48,74,79,81,82].

In the case of the multicomponent polyblends/alloys the "experiments-with-mixtures" procedure is a good starting point at least for locating the synergistic composition(s) [48] of the characteristic targeted. In the "targeted selection" procedure the success depends on the quality of a comprehensive database in which realistic *ranges* of the material functions' parameters are incorporated; the ranges should reflect the batch-to-batch variation and repeatability of the compounding process.

ACKNOWLEDGEMENT

Concepts developed here follow fruitful interaction with graduate students taking the Polyblends/Alloys Course [35]. Especially solutions developed within course assignments/projects by Drs. S. K. Dey, S. T. Lee and Mr. B. H. Luor are of particular merit.

SYMBOLS AND DESIGNATIONS

A = interface area (domain of volume v/matrix) in the laminar mixing of PBs, PABs, μm^2

$<a>$ = domain (minor phase) mean area, μm^2

BD = batch distributive (Banbury, CM, plastograph)

Br = Brinkman number, dimensionless ratio of heat generation and transport

C_i = mass fraction of the minor component

CD = continuous distributive (e.g., static) mixer

$<d>$ = mean domain size (e.g., length, diameter, thickness), μm

d = die (capillary) I.D., m

D = diffusivity const., m^2/sec

De = Deborah number (dimensionless elasticity criterion), t^*/τ_{ch}

DIP = dispersive mixer

DIT = distributive mixer

e = exponent in Equation (A.3) $[= (m/n)_i - (m/n)_m]$ or, in Equation (9), $= q_m \cdot k_i - k_m$

ER = elasticity ratio $[N_{1r}, (G')_r,$ etc.)

FR = melt flow rate (ASTM D1238), gms/600 secs

g = ratio of the inherent shear rates, $\dot{V_i}/\dot{V_m}$ [Equation (9), $= q_m \cdot k_i - k_m)$

H_1 = relaxation time spectrum, 1st approximation

G' = storage modulus, kPa

i = subscript "inclusion" (minor component), or "isotactic" as in i-PP

IS = interdomain spacing, intensity of segregation or texture in PB/PAB

ISR = inherent shear rate, sec^{-1} (i.e., $\dot{V_i}$ at a p_{12} level, which is proportional to η_i, $\dot{V_i} \equiv \eta_i \cdot p_{12}$)

k = elasticity exponent [Equation (6): $\psi_1 = \psi_1^0 \cdot s^k$; $= 1/m)$ see m

l = screw pitch or capillary length, m

L = screw length, m

LF = land fracture, instability of melt flow (at wall)

m = subscript "matrix" (major component) or power law (elasticity) exponent

MEI = mixing effectiveness indicator [Equation (33)]

MER = melt elasticity ratio $(\psi_1)_r$

MF = melt fracture, instability of flow at entrance

MMI = extrudate swelling during FR measurements (Melt Memory Index)

n = power law (viscosity) exponent

N_1 = principal normal stress difference, kPa

0 = superscript "standard"

p_{ij} = stress tensor component (e.g., p_{12} = shear)

PAB = polyalloy (polyblend + compatibilizer)

PB = polyblend

PRC = particular rheology composition

Pe = Peclet number (dimensionless mixing criterion)

q = $1/n$, reciprocal of the power law exponent; index "component q" in a polyblend [Equation (28)]

r = subscript "ratio" (as in N_{1r}) or radius of a capillary

RTD = residence time distribution

S = p_{12}/η^0

SSE = single screw extruder

t = time, secs

T = temperature, K

t^* = characteristic time of a process, secs

TSE = twin screw extruder

$\dot{V}$ = apparent shear rate, sec^{-1}

v = volume

v_i = volume fraction of the minor phase

VR = viscosity ratio (η_r)

v_T = specific volume at the temperature T, cm$^3 \cdot$ g^{-1}

w = mass fraction (see C_i, x)

ZSK = DIP, TSE compounder

x = molar fraction (see C_i, x)

β = extrudate swelling function (see MMI)

η^*, η^0 = shear viscosity (complex; standardized [17], respectively), kPa$\cdot$sec

γ = strain (shear)

γ^+ = total strain

$\dot{\gamma}$ = shear rate, sec^{-1}

Δ = difference in a characteristic (ΔN_1, $\Delta\dot{V}$, etc.) or amplitude of the process parameter ($\Delta\dot{Q}$, ΔT_m, etc.)

λ = elongation viscosity, kPa$\cdot$sec

τ_{ch} = characteristic material time (e.g., relaxation time), a moment of the relaxation spectrum, secs

χ = constant in the Equation (A.3) $[= N_{1r}^0 \cdot (\eta_i^0)]^{x_i}$ or in Equation (9) $[(\psi_1^0)_r \cdot g]$

Φ = coefficient, Equation (33), see MEI, pumping efficiency (drag flow/pressure flow)

ψ_1 = principal elasticity coefficient ($= N_1/\dot{\gamma}^2$)

x = kinematic viscosity $[= \eta \cdot V_T$(m$\cdot$sec) Equation (30)]; compressibility

II_d = time averaged 2nd invariant of the deformation rate tensor

ϕ = fluidity, $[= 0.1$ (kPa)$^{-1}]$

ν = interfacial tension, N$\cdot$m

æ = Poisson ratio

REFERENCES

1. (a) Paul, D. R. and S. Newman, eds. *Polymer Blends, Vol. 2*. Academic Press, Chapter 21 (1978); (b) Paul, D. R. *Vol. 2*. Chapter 16.

2. Plochocki, A. P. and J. L. White. *Proc. VIth Intl. Congr. Rheol.* Gothenburg, p. 522 (1976).

3. Plochocki, A. P. *Polymer Eng. Sci.*, (a) 22:1153–65 (1982); (b) 23:612 (1983); (c) 26:82 (1986).

4. (a) Cogswell, F. N. *Polymer Melt Rheology*. Halsted/Wiley (1981);(b) Middleman, S. *Flow of High Polymers*. Interscience (1968); (c) Greassley, W. W. *Adv. Polymer Sci.*, 16:1–230 (1974).

5. Utracki, L. A. *Polymer Eng. Sci.*, 23:446 (1983) and [41]; Sanchez, I. C. *Ann. Rev. Mater. Sci.*, 5:387 (1968).

6. Litt, M. H. *Trans. Soc. Rheol.*, 20:47 (1976).

7. (a) Semjonov., V. *Adv. Polymer Sci.*, 5:387 (1968); (b) Westover, R. F. *SPE Transactions*, 1:1 (1961).

8. Danesi, S. and R. S. Porter. *SPE Tech. Papers*, TP24:240 (1976); and *Polymer*, London, 19:448 (1978).

9. White, J. L. et al. *J. Appl. Polymer Sci.*, 16:1313 (1972).

10. Han, C. D. (a) and T. C. Yu. *J. Appl. Polymer Sci.*, 15:1163 (1971); (b) *Rheology of Polymer Processing*. Wiley (1976).

11. Doppert, H. L. and W. S. Overdiep. in "Multicomponent Polymer Systems," ACS *Adv. Chem. Series*, 99:53–106 (1971).

12. e.g., H. P. Schreiber and S. H. Storey. *J. Polymer Sci.*, letters, 3B:724 (1965).

13. e.g., L. P. McMasters in "Copolymers, Polyblends and Composites," *ACS Adv. Chem. Series*, 142:43–65 (1975).

14. Lyngaae-Jorgensen, J. *ACS Org. Coat. & Polymer Div.*, Preprints, 45:1979 (1981).

15. VanOene, H. (a) in [1] *Vol. 1*, Chapter 7; (b) with Z. Mencik and H. K. Plummer. *J. Polymer Sci.*, 10A-2, 507 (1972).

16. Miller, S. Preprints PRI Intl. Conf., "Toughening Plastics," London, paper #8 (1978).

17. Kelvey, J. M. *Polymer Processing*. Wiley (1962).

18. Min, K., J. L. White and J. F. Fellers. *Polymer Eng. Sci.*, 24:1327 (1984).

19. Plochocki, A. P. *Trans. Soc. Rheol.*, 20:287 (1976).

20. Martinez, C. B. and M. C. Williams. *J. Rheol.*, 24:421–50 (1980).

21. Han, C. D. *Multiphase Flow in Polymer Processing*. Academic Press (1981).

22. Arrhenius, S. *Zeitschr. Phys. Chem.*, 1:285 (1887); and C. Lees. *Proc. Phys. Soc.*, 17:460 (1900); *Phil. Mag.*, 16:128 (1901).

23. Bogue, D. C., T. Masuda, Y. Einaga and S. Onogi. *Polymer* (Kyoto), 1:563 (1970).

24. Racin, R. and D. C. Bogue. *J. Rheol.*, 23:263 (1979).

25. Carley, J. F. (a) and S. C. Crossan. *Polymer Eng. Sci.*, 21:249 (1981), and (b) *SPE Tech. Papers*, 30:439–42 (1984).

26. Kasajima, M. and Y. Mori. *Kagaku Kogaku*. 37:915 (1973); M. Kasajima et al. *Trans. (Jap) Soc. Rheol.*, 7:27 (1979); *Proc. Intl. Conf. Polymer Processing*, Cambridge, MA:MIT, pp. 508–27 (1980).

27. Bendaikha, H. "LLDPE Polyblends—A Simple Rheological Model for Morphology Formation in Melt Flow," MSci Thesis, ChE, SIT (1987).

28. Lee, B. L. and J. L. White. *Trans. Soc. Rheol.*, 19:481–91 (1975).

29. (a) Kuleznev, V. N. et al. *Koll. Zh.*, 27:459 (1965); (b) *Smesi Polimerov*. Khimya, Moskva (1980).

30. Dunlop, A. N. and H. L. Williams. *J. Appl. Polymer Sci.*, 17:2945 (1973).

31. E. C. Bingham. *Fluidity and Plasticity*. McGraw-Hill (1922).

32. Hayashida, K., J. Takahashi and M. Matsui. *Proc. 5th Intl. Congr. Rheol.*, 4:525 (1970).

33. Karian, H. G. et al. "Mixing Studies of a Polyalloy," *Polymer Eng. Sci.*, in print (1989).

34. Plochocki, A. P. and H. Emmanuilidis. *ACS Polymer Preprints*, 26(1):288 (1985).

35. Plochocki, A. P. Notes for course "Polymer Blends, Alloys and Composites: (ChE 235), ChE Dept., Stevens Institute of Technology (1983/1984).

36. Cogswell, F. N. and D. E. Hanson. *Polymer* (London), 16:936–7 (1975).

37. Freeguard, G. F. and M. Kamarkar. *J. Appl. Polymer Sci.*, 15:1649–55 (1971).

38. Galeski, A. et al. *J. Polymer Sci.*, 22(A-2):739 (1984).

39. Naar, R. Z. and R. F. Heitmiller. *Proc. BSR-PRI Conference on Plastics.* London, pp. 65–80 (1962).
40. Sjordsma, S. D., A. C. A. M. Bleijenberg and D. Heikens. *Polymer,* London, 22:619–24 (1981).
41. Utracki, L. A. *J. Macromol. Sci.,* B18:731 (1980).
42. Prest, W. M. and R. S. Porter. *J. Polymer Sci.,* 10A-2:1639 (1972).
43. Avgeropoulos, G. N. et al. *Rubber Chem. Technol.,* 49:93 (1976).
44. Pakula, T. et al. *Polymer Bull.,* 2:799–804 (1980).
45. Squires, P. H. Materials PIA Course #190 "Quality in the Extrusion Process," Stevens Institute of Technology, Castle Point, NJ 07030 (Dec. 1982).
46. Faucher, J. A. *J. Polymer Sci.,* 12A-2:2153–5 (1974).
47. White, J. L. and A. P. Plochocki. "Correlation of Measured and Estimated Normal Stress Differences as a Test of Continuum-Rheology Approach for Polyblends," unpublished (1976).
48. Cornell, J. A. *Experiments with Mixtures.* NY:Wiley (1982).
49. Andrews, R. D. et al. "The Interface in Binary Mixtures of Polymers," Ann. Mtg. AIChE, NYC, paper #41 (Nov. 1987).
50. Plochocki, A. P. et al. *ACS-PSME Preprints,* 54:245 (1986).
51. Elmendorp, J. J. *Polymer Eng. Sci.,* 26:415 (1986).
52. Shundo, M. *J. Appl. Polym. Sci.,* 10:939 (1966).
53. Oyanagi, Y. *Kobunshi Ronbunshu,* 32:348 (1975).
54. Satake, K. *J. Appl. Polym. Sci.,* 15:819 (1971).
55. Plochocki, A. P. *Adv. Polym. Technol.,* 3:405 (1984).
56. White, J. L. *Nihon Rioriji Gakkaishi,* 12:184 (1984); *Polymer Eng. Sci.,* 19:818 (1979).
57. Pakula, T. "A Flow-Through Mixer," Pat. PL103982 (Feb. 21, 1976) and in *Polymer Blends.* Martuscelli, E. et al., eds. Plenum (1982).
58. Clegg, P. L. *Pure Appl. Chem.* 55:755–64 (1983).
59. Heikens, D. *Kem. Ind.,* 31:165 (1982).
60. Fayt, R. et al. *J. Polym. Sci.,* 20(A-2):2209 (1982).
61. Wen, A. "PE/PS Polyblends: Viscosity Ratio Determination for Commercial Grades and CFD Morphology Determination," Special Ch. E. Project, Stevens, A. P. Plochocki, advisor (May 1984).
62. Elmendorp, J. J. and R. J. Maalcke. In *Interrelations between Processing, Structure and Properties of Polymeric Materials.* J. C. Seferis and P. S. Theocaris, eds. Elsevier, pp. 219–30 (1984); *Polym. Eng. Sci.,* 25 (1985).
63. Jevanoff, A., E. N. Kresge and L. L. Ban. "Morphology of Thermoplastic Polyolefins Blends by SEM," *PLASTICON 81, Polymer Blends, PRI Conf., Proc.,* paper #23.1-14 (Sept. 1981).
64. Hong. *Macromolecules,* 17:1531–7 (1984); and *Polym. Bull.,* 7:561–6 (1982).
65. Manas-Zloczower, I., A. Nir and Z. Tadmor. *Rubber Chem. Technol.,* 57:583 (1984).
66. Dreval, V. E. et al., *Rheol. Acta.,* 22:102–7 (1983).
67. Plochocki, A. P. *Industrial Polymer Blends: Engineering & Technology.* Hanser, in preparation (1989).
68. Tadmor, Z. and C. G. Gogos. *Principles of Polymer Processing.* Wiley, Ch. 7, 11 (1979).
69. Hold, P. *Adv. Polym. Technol..* 2:197–228 (1982); and *Kunststoffe,* 34:1027 (1981).
70. Plochocki, A. P., S. K. Dey and A. Wilczynski. *Polym. Eng. Sci.,* 26:1007 (1986).
71. Powell, R. L. and S. G. Mason. "Dispersion by Laminar Flow," *AIChE J.,* 28:286–93 (1982).
72. Thorn, A. D. *Thermoplastic Elastomers.* RAPRA, Shawbury, U.K. (1980).
73. Zahorski, S. *Mechanics of Viscoelastic Fluids.* Boston:Kluver (1982).
74. Derringer, H. C. G. *J. Appl. Polym. Sci.,* 18:1083 (1976).
75. Utracki, L. A. In *Current Topics in Polymer Science.* S. Inoue et al., eds. Hanser (1986).

76. Griskey, R. G. and K. Maciejewski. *SPE TP20*, 23–7 (1974).

77. Dey, S. K., A. Kiani, A. P. Plochocki and J. E. Curry. *Kunststoffe—German Plastics,* 76:455 (1986).

78. Plochocki, A. P. and J. L. White. "Rheology and Morphology of the Rubber/Polypropylene Blends," Univ. Tennesse at Knoxville, PATRA #191 report (July 1982).

79. Plochocki, A. P. and G. Miller. *Chemical Industry,* Warsaw, 55:539 (1976).

80. See, e.g., Mack M. et al. *Kunststoffe/German Plastics*, 3:257/30 (1988): Curry, J. E. et al. *J. Elastomers and Plastics*, 18:256 (1986).

81. Sebastian, D. H. and J. A. Bresenberger. *SPE TP29*, 121 (1983).

82. Malinowski, E. R. and D. R. Howery. *Factor Analysis in Chemistry*, Wiley (1980).

Milton Keynes UK
Ingram Content Group UK Ltd.
UKHW040058071024
449327UK00019B/642